P9-CKX-654

recent advances in phytochemistry

volume 17

Mobilization of Reserves in Germination

RECENT ADVANCES IN PHYTOCHEMISTRY

Proceedings of the Phytochemical Society of North America
General Editor: Frank A. Loewus *Washington State University, Pullman, Washington*

Recent Volumes in the Series

Volume 11 **The Structure, Biosynthesis, and Degradation of Wood**
Proceedings of the Sixteenth Annual Meeting of the Phytochemical Society of North America, Vancouver, Canada, August, 1976

Volume 12 **Biochemistry of Plant Phenolics**
Proceedings of the Joint Symposium of the Phytochemical Society of Europe and the Phytochemical Society of North America, Ghent, Belgium, August, 1977

Volume 13 **Topics in the Biochemistry of Natural Products**
Proceedings of the First Joint Meeting of the American Society of Pharmacognosy and the Phytochemical Society of North America, Stillwater, Oklahoma, August, 1978

Volume 14 **The Resource Potential in Phytochemistry**
Proceedings of the Nineteenth Annual Meeting of the Phytochemical Society of North America, Dekalb, Illinois, August, 1979

Volume 15 **The Phytochemistry of Cell Recognition and Cell Surface Interactions**
Proceedings of the First Joint Meeting of the Phytochemical Society of North America and the American Society of Plant Physiologists, Pullman, Washington, August, 1980

Volume 16 **Cellular and Subcellular Localization in Plant Metabolism**
Proceedings of the Twenty-first Annual Meeting of the Phytochemical Society of North America, Ithaca, New York, August, 1981

Volume 17 **Mobilization of Reserves in Germination**
Proceedings of the Twenty-second Annual Meeting of the Phytochemical Society of North America, Ottawa, Ontario, Canada, August, 1982

---

recent advances in phytochemistry

volume 17

Mobilization of Reserves in Germination

Edited by

Constance Nozzolillo

University of Ottawa
Ottawa, Ontario, Canada

Peter J. Lea

Rothamsted Experimental Station
Harpenden, Herts, United Kingdom

and

Frank A. Loewus

Washington State University
Pullman, Washington

PLENUM PRESS • NEW YORK AND LONDON

Library of Congress Cataloging in Publication Data

Phytochemical Society of North America. Meeting (22nd: 1982: University of Ottawa)
Mobilization of reserves in germination.

(Recent advances in phytochemistry; v. 17)
"Proceedings of the annual symposium of the Phytochemical Society of North America...held August 2–6, 1982, at the University of Ottawa, Ottawa, Ontario, Canada"—Verso t.p.
Bibliography: p.
Includes index.
1. Germination—Congresses. 2. Plants—Metabolism—Congresses. 3. Plant translocation—Congresses. 4. Plant membranes—Congresses. I. Nozzolillo, Constance. II. Lea, Peter J. III. Loewus, Frank Abel, 1919– . IV. Title. V. Series.
QK861.R38 vol. 17 [QK740] 582′.0333 83-9569
ISBN 0-306-41377-9

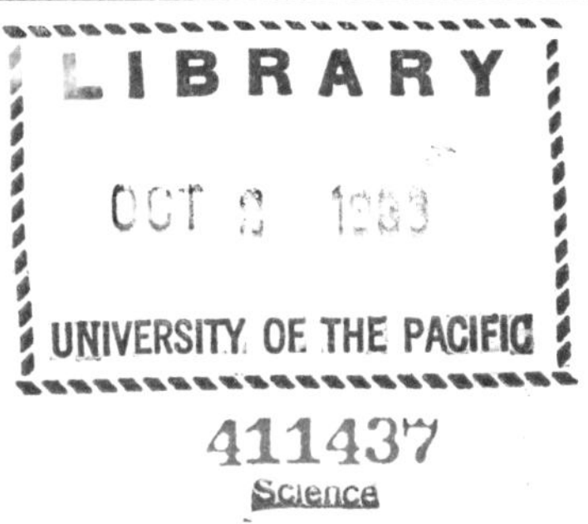

Proceedings of the Annual Symposium of the Phytochemical Society of North America on Mobilization of Reserves in Germination, held August 2–6, 1982, at the University of Ottawa, Ottawa, Ontario, Canada

A Division of Plenum Publishing Corporation
233 Spring Street, New York, N.Y. 10013

Printed in the United States of America

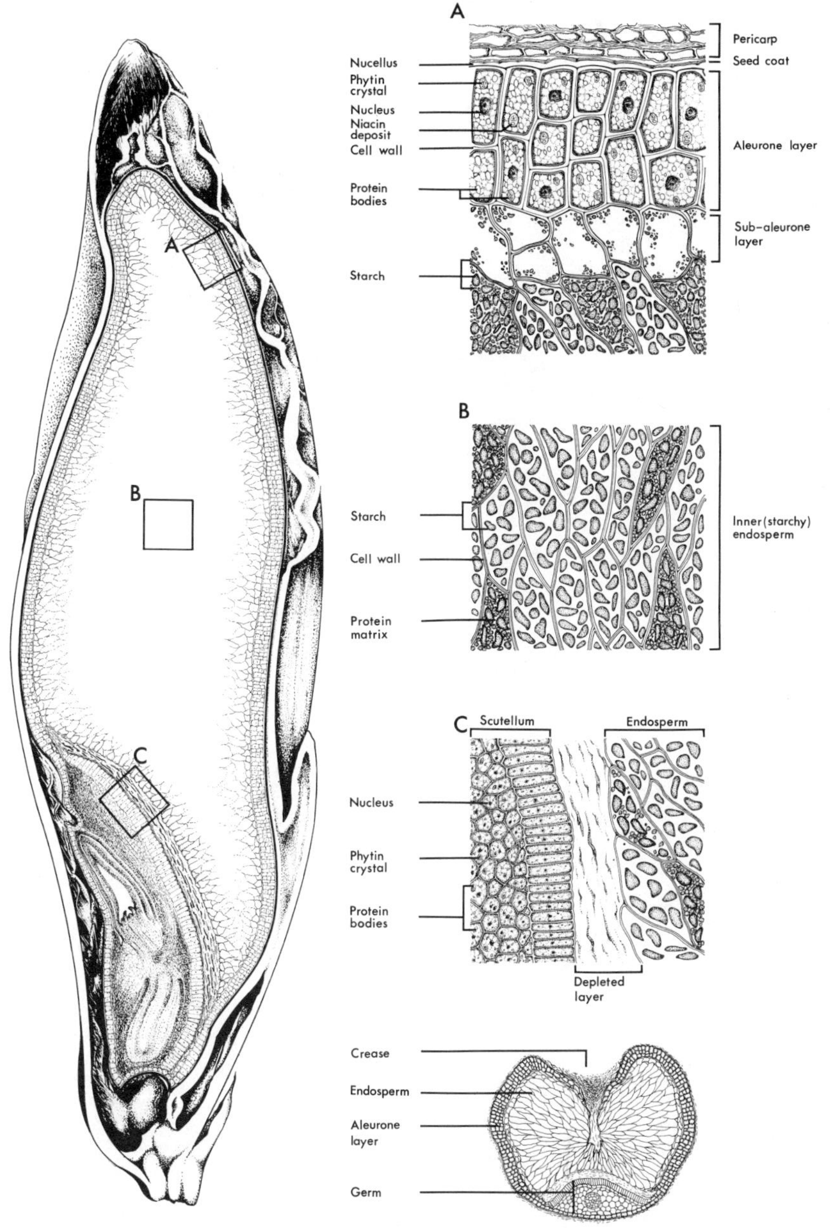

Diagram of the anatomy of a barley (Hordeum vulgare L.) kernel showing the tissue relationships in longitudinal (left) and cross section (lower right). Enlargements show cellular details in the bran (A), central endosperm (B) and at the embryo-endosperm interface (C). Diagram provided by R.G. Fulcher and S.I. Wong, Ottawa Research Station, Agriculture Canada.

PREFACE

Although many reviews and articles about germination have been published, our understanding of the process is far from complete. Some of the reactions involved in the transformations and translocation of reserve compounds and their final fate in the growing tissues was the subject of the annual symposium of the Phytochemical Society of North America which was held in August, 1982, on the campus of the University of Ottawa, Canada. A major emphasis was on low molecular weight compounds, amino acids and sugars, and the checks and balances operating as they are produced by hydrolysis in reserve tissues and transformed in the growing tissues. The critical role of membranes was given especial emphasis. The closing presentation was an anthropocentric review of the nutritional benefits accrued during germination, and provided a scientific basis for the inclusion of sprouted seedlings in the human diet. Dr. R. G. Fulcher introduced the symposium topic with an elegant histochemical study in which the site and nature of seed reserves were defined by means of the glowing colors of fluorescence microscopy.

The symposium was organized by C. Nozzolillo with advice and assistance from fellow PSNA members S. A. Brown, J. T. Arnason, and A. Picman and a visiting scientist from Sweden, C. Liljenberg.

The PSNA itself underwrote a considerable part of the cost of the symposium, but the added support of a conference grant from the National Science and Engineering Research Council of Canada, a travel grant from Queen's University of Belfast to Dr. E. Simon, and provision of facilities by the University of Ottawa, in particular the able assistance of M. Olivier Bilodeau, Administrative Assistant to the Dean of the Faculty of Science and Engineering, is greatly appreciated.

Although the symposium proper was not in the usual purview of secondary metabolites with which the Society is most often concerned, the after-dinner talk by Dr. G. A. Rosenthal was a recognition of the protective role these seemingly useless compounds play in seeds and in the rapidly growing field of phytochemical ecology in general.

Toni Smith assisted in proof-reading and produced the word-processed final copy that was used for camera-ready production. Her excellent help is deeply appreciated.

January, 1983

C. Nozzolillo
P. J. Lea
F. A. Loewus

CONTENTS

Chapter One

INTRODUCTORY CHAPTER

CONSTANCE NOZZOLILLO

Department of Biology
University of Ottawa
Ottawa, Ontario
Canada K1N 9B4

Plants produce a wide variety of reproductive propagules designed to enable them to survive periodic harsh climatic conditions and to increase and multiply thereafter. The process of resurgence of growth of these small packages of genetic information is called germination. In the present volume, germination of only one kind of these propagules is considered: that of the seed or one-seeded fruit of flowering plants. Seeds can contain large amounts of food reserves and for this reason are of fundamental importance to human society. Cultivation of seed-producing plants for purposes of human consumption is the basis of agricultural activities, and the primary reason for the enduring interest shown by humans in the germination process. Germination of seeds is not a simple process and depends on many factors. Thus it cannot be effectively covered by a volume such as this one and a further delimitation of the topic to the interim period between the onset of germination, signaled by imbibition or uptake of water by the seed, and development of the independent plantlet, is made. The fundamental importance of the entire process is indicated by the several books that have been written about it, most recently, for example, those by Khan,[1] Mayer and Poljakoff-Mayber,[2] and Bewley and Black.[3]

The first step a reviving seed has to take is to hydrate its tissues and transform its membranes from their dormant resting state to an active state. The mechanism by which this is brought about is not fully understood and two major theses have been proposed. Both views are presented in this volume. In Chapter 2, Simon and Mills propose an interim "leaky" state as a result of rearrangement of phospholipids into a hexagonal phase. McKersie and Senaratna in Chapter 3 uphold the widely accepted view that no fundamental change in membrane arrangement occurs and that "leakiness" can be explained by physical chemistry.

Mobilization is a general term covering the ensuing events. It includes both breakdown of reserves and their movement or translocation to the growing parts of the young plant. The reserves in question are of three major types: carbohydrates, proteins and lipids, and the various kinds of seeds vary widely in the composition and location of reserves. Most of the agriculturally important species, less that 0.01% of the total number of known plant species, belong to two families representative of the two major taxonomic divisions of the flowering plants: the dicotyledonous Fabaceae (Leguminosae) and the monocotyledonous Poaceae (Gramineae). These are more commonly known by the names of pulses or grain legumes and cereal grains. The reserves of grain legumes, peas, beans and the like, are stored in the embryo cotyledons and have a relatively high protein content. Reserves of cereal grains are found in the endosperm, a tissue characteristic only of the flowering plants, and are predominantly carbohydrate in nature. Because of their importance to the baking and brewing industries, the chemistry and histochemistry of cereal grains has been much more intensively studied than that of the grain legumes. The typical structure of a cereal grain is exemplified by the drawing of a barley grain in the frontispiece. Identification and localization of the various components has been made through the combined use of chemical analysis of isolated fractions and *in situ* histochemistry, in particular, the sensitive technique of fluorescence microscopy.[4] Additional minor constituents of the bran fraction revealed by this method but not indicated in the frontispiece are phenolic acids and aromatic amines which may act as deterrents to insect or microbial attack and/or as germination modifiers. Similar studies of dicotyledonous seeds are now being published.[5]

Nitrogen reserves of the seed are of critical importance to their usefulness as human food. The presence of comparatively high protein reserves in legumes is what makes them such a valued addition to a diet based on cereal grains as a staple. Nitrogen is also an essential part of the enzymes that appear during germination, either via *de novo* synthesis or by activation of previously stored complexes and as such touches on virtually all aspects of seed metabolism. The study of these enzymes and of the processes by which they make their appearance is not a major consideration in the present volume. Protein hydrolyzing enzymes are discussed in a recent symposium volume[6] but less attention has been paid to the fate of the hydrolysis products. It is likely that most of the amino acids are not reutilized in the form in which they are released from the reserves. The authors of Chapters 4 and 5 have made notable contributions to this important topic. Oaks presents an in depth investigation of the control of amino acids over their own fate in the developing corn seedling, and Lea and Joy discuss the form in which nitrogen reserves are moved through the pea seedling. Nitrogen reserves of the seed also include those vital molecules that are the basis of cellular information and control: the nucleoproteins. These substances and the various nucleic acids, nucleotides and other purine and pyrimidine components of the cell, are not discussed in the present volume, but Osborne[7] and Sussex[8] in a recent symposium give them due recognition as the very basis of transmittal of information from one generation of plants to another. A variety of low molecular weight nitrogenous compounds of no evident function may form a substantial proportion of total nitrogen reserves. *Theobroma cacoa* and *Coffea* species, for example, have been cultivated for centuries for the non-protein nitrogen content of their seeds, but for what reason does the plant deposit such substances there and what is their fate during germination? The answer to both these questions is beyond the scope of this volume, but is found in modern ecological studies of the chemical means by which plants defend themselves.[9]

In many grain legumes and all cereal grains, starch is a major reserve material. The enzyme activities reawakened to hydrolyze starch are the amylases. Because of the importance of these enzymes to the brewing industry, they are probably among the most thoroughly studied of all the enzymes in the germinating seed.[6] Much less work has been done on the products of hydrolysis, and in particular, on the molecule which serves as the primary form for movement of carbon reserves from the storage tissues. In the growing plantlet, the cell wall is a major destination of carbon not used for energy production. That is, one glucose polymer, starch, is transformed into another, cellulose. But glucose is not the molecule transported. The role of sucrose, a disaccharide composed of glucose and fructose, as the major translocate of the carbon chains of the plant is well-known.[2] The innovative study reported by Maclachlan and Singh in Chapter 8 provides a detailed look at the fate of translocated sucrose in the developing pea seedling.

Starch is not the only form of carbohydrate reserves. Many oil-bearing seeds contain unusual oligosaccharides that are of chemotaxonomic interest, for exmple, umbelliferose in seeds of the Umbellifereae, and planteose in seeds of the Plantaginaceae.[10] Cell wall material ("hemicelluloses") may also serve as reserves digested during germination: galactomannans of the endospermic legumes such as Trigonella species[1] and mannans in umbelliferous species.[10] Bewley, Leung and Quellette in Chapter 7 present the results of recent studies on mobilization of endospermic mannan reserves of lettuce seeds, including an examination of the specific enzyme activities called into play.

One of the triumphs of the past fifteen years in our understanding of plant metabolism has been the discovery of the microbodies (glyoxysomes) that appear during germination of oil-bearing seeds.[11] In this volume we have chosen to look at lipids only in their function as an integral component of the membranes (Chapters 2 and 3) and as a particular kind of complex with carbohydrates. The starch grain is a highly ordered structure (amyloplastid) whose shape is characteristic of the species producing it. It contains various minor components in addition to the amylose and amylopectin backbone. The present state of knowledge of some of these minor constituents, in particular starch-bound monoacyl lipids, is reviewed by Galliard in Chapter 6.

The major reserves of the seed have received the most attention in scientific studies, but seeds also contain other equally important constitutents even though in lesser amounts. Among these are the mineral reserves.[12] Studies of the mobilization of these reserves are limited in number. The bulk of the macro elements, K, Mg, Ca, are buried in protein bodies in association with phytin. Phytin is also the major reserve of phosphate essential to the energy-producing machinery of the cell and its hydrolysis by phytase has been well studied in that context.[2] What little information we have on the fate of the much less-studied myo-inositol moiety is reviewed in this volume by Loewus in Chapter 9.

Control of the processes of mobilization is complex, and depends on factors internal and external to the seed. The roles of light, temperature, and humidity are not a consideration of this volume but considerable attention is paid to internal control mechanisms. Among these, plant hormones are the best known, and least understood. Certainly there is an interaction of nucleic acids and hormones,[7,8] and evidence is mounting that membranes are a major target of hormone action.[1] As a result of the enormous strides that have been made in analysis[13] of the minute quantities of hormones typically present, our understanding of the mode and site of action of these compounds has considerably increased, but much still remains to be learned. A complicating factor is the existence of hormones in biologically active and inactive forms. The mode of transport of the various forms of auxin from one part of the corn seedling to another is presented by Bandurski in Chapter 11, and a summary of recent studies of gibberellin activity in the germinating seed, particularly as it affects membrane permeability, is presented by Black, Chapman and Norman in Chapter 10.

In addition to protein, carbohydrate, and lipid content of seeds, the fate and role of vitamins are of considerable interest, from the viewpoint of human nutrition if for no other reason. They were a fashionable area of study a generation ago.[14] A renewed scientific interest in these components has paralleled the modern-day concern of the "natural food" proponents that essential elements may be missing from processed foods. The desirability of including sprouted seedlings in the diet was realized by the Chinese long before vitamins were discovered. Finney's review in Chapter 12 of the nutritional value of sprouted seedlings is a fitting end to this volume. A fitting ending to this introductory chapter is a reminder to the reader that the work done to date on germination, although impressive and extensive, is little more than an auspicious beginning. We have come to understand more completely the workings of a few economically important species as their seeds spring to life from the resting state, but thousands of wild species remain to challenge future generations of investigators. To date much of the interest in wild species has concerned their remarkable adaptation to the environment in their control of the dormant state. The distressing success of many noxious weed seeds rests in their ability to resist germination at a time when agricultural practices would result in their destruction. Study of the chemistry and mobilization of reserves of wild species of little or no interest to agriculture, silviculture, or horticulture, has only begun. A preliminary study[15] of two such species, *Impatiens capensis* and I. pallida, has revealed many unusual features of the seeds of these common North American plants, including the unexpected presence of planteose (Kandler, personal communication).

REFERENCES

1. KHAN AA, (ed) 1977 The Physiology and Biochemistry of Seed Dormancy and Germination. North Holland, Amsterdam New York
2. MAYER AM, A POLJAKOFF-MAYBER 1982 Germination of Seeds. 3rd edit, Pergamon, London
3. BEWLEY JD, M BLACK 1978 Physiology and Biochemistry of Seeds in Relation to Germination. Vol 1. Development, Germination and Growth. Springer, Berlin Heidelberg New York
4. FULCHER RG 1982 Fluorescence microscopy of cereals. Food Microstruct 1: 167-175
5. YIU SH, H POON, RG FULCHER, I ALTOSAAR 1982 The microscopic structure and chemistry of rapeseed and its products. Food Microstruct 1: 135-143
6. VAUGHAN JG, J MASSE, J DAUSSANT (Eds) 1982 Seed Proteins. Academic, London New York
7. OSBORNE DJ 1983 Control systems operations in the early hours of germination: a biochemical study in the Graminae. Can J Bot 61: in press
8. SUSSEX IM 1983 The regulation of synthesis of embryo-specific proteins in plants. Can J Bot 61: in press
9. ROSENTHAL GA 1982 Plant Nonprotein Amino and Imino Acids: Biological, Biochemical, and Toxicological Properties. Academic, New York London
10. KANDLER O, H HOPF 1980 Occurrence, metabolism, and function of oligosaccharides. In J Preiss, ed, Carbohydrates: Structure and Function, The Biochemistry of Plants, Vol 3, Academic, London New York
11. BEEVERS H 1979 Microbodies in higher plants. Annu Rev Plant Physiol 30: 159-193
12. LOTT JN, JS GREENWOOD, CM VOLLMER 1982 Mineral reserves of castor beans: the dry seed. Plant Physiol 69: 829-833
13. BRENNER M 1981 Modern methods for plant growth substance analysis. Annu Rev Plant Physiol 32: 511-538
14. CROCKER W, LV BARTON 1957 Physiology of Seeds. Chronica Botanica, Waltham, Mass

15. NOZZOLILLO C 1982 Mobilization of reserves during germination of seeds of *Impatiens capensis* Meerb. and *I. pallida* Nuttall. Abstracts 12th Intern Congr Biochem, Perth, Australia

Chapter Two

IMBIBITION, LEAKAGE AND MEMBRANES

ERIC W. SIMON AND LORNA K. MILLS

Department of Botany
The Queen's University of Belfast
Northern Ireland

INTRODUCTION

The process of imbibition is accompanied by a variety of other events--the activation of enzyme systems, the start of active metabolism and the leakage of solutes. The intention of this chapter is to present the facts about water uptake and leakage briefly, and then attempt a critical review of some of the interpretations that have been proposed.

The leakage of solutes is not unique to seeds, for it occurs also when dry spores, pollen, bryophytes, or nematodes imbibe water. However as this volume is focused on seed germination, reference will only be made to what is known of these other systems if it helps to elucidate what happens in seeds.

IMBIBITION

Air-dry seeds generally contain about 10 to 15% water, but given the opportunity they readily imbibe water because of their very low water potential, which is in the order of -1000 bars.[1]

The rate at which seeds imbibe water depends in the first place on the gradient of water potential between water and seed. Imbibition will be fastest when dry seeds with very low water potential are placed in pure water. As the seed takes up water and its water potential rises gradually towards zero, the gradient will decline and imbibition will slow down in consequence. If the seeds have already been moistened before they are placed in water, for instance by equilibration with moist air, the initial potential gradient between seed and water will be relatively low, and imbibition will therefore start off at a reduced rate; the same will happen if water is supplied at a low potential, either from a dry soil or from a concentrated solution.

In addition to such factors influencing the gradient of water potential, the rate of imbibition also depends on the resistance to water flow. Although pea seeds can absorb moisture from damp air, imbibition is much more rapid in the liquid phase, being completed in about a day when peas are placed on wet filter paper, and in about 12 h when they are immersed in water. Under soil conditions, the degree of contact between seed and soil particles has a major bearing on the rate of imbibition.[2]

Once water begins to enter a seed, a second resistance to water flow comes into play. This concerns the ease with which water can penetrate through the seed tissues. If the resistance to water flow remained the same no matter how wet the tissues were, then one would expect a uniform gradient of water content in an imbibing seed, wettest at the outside and driest in the center. If on the other hand water can flow through wet tissues much more readily than through dry tissues, then the water content profile would show an abrupt transition between wet and dry regions. In fact there is just such an abrupt boundary between the outer, wet region of pea cotyledons and the dry central regions.[3] As imbibition progresses the wetting front moves inwards and the region already wetted is raised to higher water contents.

The net result of the changing pattern of water potential gradient and resistance is seen in Fig. 1. When isolated embryos are placed in water imbibition is rapid over the first 30 min and then proceeds more slowly.[3] The presence of an intact testa introduces an initial lag phase during which the testa itself becomes wetted; the testa also reduces the subsequent rate of imbibition.

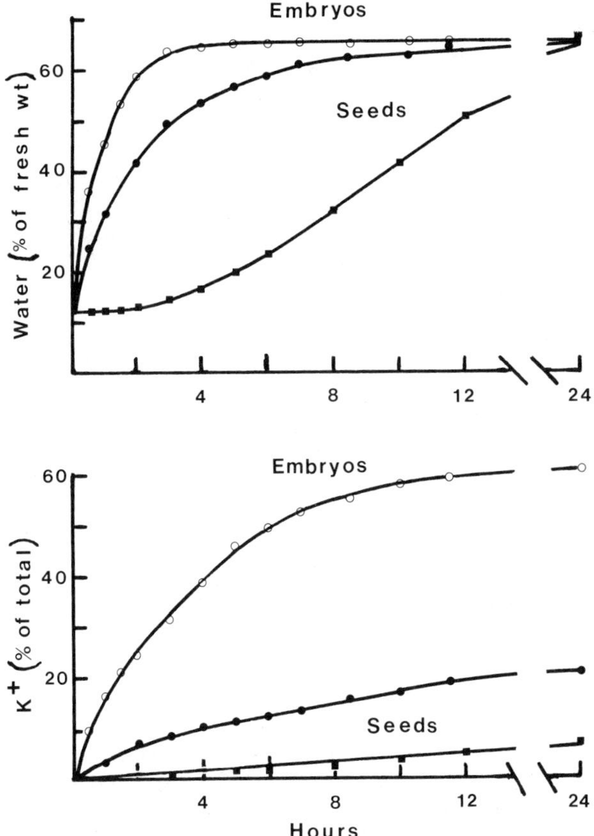

Fig. 1. Imbibition of water (above) and leakage of potassium (below) from 50 pea seeds or embryos, cv Kelvedon Wonder placed in 100 ml water. The amount of potassium leaking out is expressed as a percentage of the total amount originally present. Two curves are shown for seeds; in each graph the lower curve was obtained with greenhouse-grown seed (2% of seeds with damaged testa[5]), and the upper curve with commercial seed (40% damaged). 1980 seed.

LEAKAGE FROM SEEDS AND EMBRYOS

When seeds are placed in water, solutes leak out of them.[4] A common experimental arrangement is to place 50 seeds in 100 ml water and measure the release of solutes by the change in conductivity of the liquid or by its potassium content. The extent of leakage depends on the state of the testa (Fig. 1). There is little leakage from intact pea seeds, but seeds in which the testa has been damaged deliberately with a fine needle or a scalpel absorb water rapidly and leak quite profusely; the same is true of seeds which have suffered mechanical damage during harvest, sorting or packing.[5,6] It is not too difficult to remove the entire seed coat from some varieties of pea without injuring the embryo inside. Such isolated embryos leak considerably more than intact seeds (Fig. 1) and for this reason are often used in studies of the mechanism of leakage.

Leakage from pea embryos is most rapid in the first moments of imbibition, thereafter slowing down minute by minute until by 10 to 20 min it reaches a steady rate which continues with little change for the next half-hour or more; leakage eventually comes to a stand-still after about a day.[7-9] A similar pattern of leakage has been observed in soybean cotyledons[10] and in scarified seeds of _Lotus corniculatus_.[11]

By the time that leakage has come to a stop pea embryos have lost a high proportion of their solutes, typically about two-thirds of the electrolytes[12] or potassium (Fig. 1) are lost from pea embryos. Pea _seeds_ leak less; on average one-third of the electrolytes and potassium, and 10% of the sugars are lost from commercial pea seeds.[8] Over a two-hour period scarified _Lotus_ seeds lose 13% of their potassium and 9% of soluble protein, but only 3.6% sugar, 3.1% phosphate and 1.4% amino acids.[11]

The catalogue of solutes leaking out of seeds and embryos includes many substances characteristic of cytoplasm--a range of amino acids and sugars, organic acids, phenolics, phosphates and gibberellic acid.[4,13] Soluble enzymes also leak into the medium in which pea and peanut embryos have been imbibing--but only in small amounts, generally no more than 1 to 2% after 6 h;[14] this should be compared with a loss of 50% of the potassium from pea embryos in the same time (Fig. 1). Imbibing bean and soybean embryos lost higher levels of enzyme (up to 6% and 17% respectively) in 6 h[14] probably because of their tendency to disintegrate during imbibition, strips of tissue splitting away from the cotyledons. "After 30 to 60 min the water around the embryos becomes turbid and chalky in color, suggesting that cellular rupture has occurred"[14] as indeed is indicated by light and scanning electron microscopy.[15]

The leakage of low molecular weight solutes from seeds and embryos is most profuse when relatively dry embryos are placed in water. Seeds or embryos that have first been allowed to take up a small amount of water from moist air or moist filter paper show a reduced rate of leakage if they are subsequently immersed in water. Likewise succulent, immature peas leak less in water than peas which have been left to ripen and dry out on the plant.[7] Leakage from peas is very largely suppressed once the embryos have reached a water content of 30% or more.[16] The corresponding threshold water content for the embryos of other species is about 15 to 35% water.[10,15]

These are the salient facts about leakage; we discuss next hypotheses about the underlying mechanism(s).

HYPOTHESES ABOUT LEAKAGE

Surface Deposits

The solutes that leak away from imbibing seeds and embryos might be simply deposits that were originally located on the surface or in the apoplast, but dissolve away during imbibition. However, this seems unlikely for a number of reasons. First, a very high proportion of the electrolytes and potassium leaks out in 24 h, more than is likely to be extracellular. Second, the wide variety of solutes leaking out suggests a cytoplasmic origin. Finally, the leakage of solutes from imbibing embryos continues even when they are transferred each minute to a fresh beaker of water, although one might reasonably expect a surface deposit to be washed away after two or three rinses in water.[8] Likewise when embryos are dried back to their initial weight after 30 min imbibition, and then returned to water the leakage of solutes continues unabated, a sequence that can be repeated through several cycles.[7]

Membrane Rupture

When water passes across a cell membrane in response to a gradient of osmotic potential of about 10 bars, the process is orderly, with no suggestion that the membranes are rendered leaky. Larson[17] proposed that when water flows into imbibing tissues it causes serious "cell membrane damage", allowing the free release of cytoplasmic solutes. As imbibition is driven by a gradient of around 1000 bars, the inrush of water may well be so violent as to disrupt membrane organization more or less completely. Such damage could presumably only affect a proportion of the cells of an embryo, for otherwise one would have to expect a complete, 100% loss of all solutes. A number of experimental approaches have been used to test this hypothesis.

Tetrazolium staining. Powell and Matthews[18] used the tetrazolium stain to investigate the state of membranes in the abaxial surface of pea cotyledons after imbibition. Cotyledons taken from intact seeds showed much more complete and uniform staining than those imbibed without a testa. Sectioning of the cotyledons from such imbibed embryos treated with tetrazolium showed that a layer of surface cells failed to stain; in peas and soybeans this unstained layer is reported to be about 0.4 mm thick.[19] We find that in peas (cv. Kelvedon Wonder) the unstained layer may be 0.6 to 0.9 mm thick depending on the batch of seed. The volume occupied by this zone of unstained cells is quite substantial, amounting to 40 to 50% of the volume of the cotyledons.

At first the unstained cells were thought[18] to be "dead", a term frequently used in the seed-test literature, and one which implies a breakdown of cell organization, due to the inrush of water; intact seeds imbibe more slowly, cell structure presumably remains intact and the cells therefore stain with tetrazolium. However it is now recognized[9] that the unstained cells are not bereft of all activity, for they stain up if imbibed in succinate solution or if imbibed in water and then treated with succinate along with the tetrazolium salt. The incomplete staining of imbibed embryos therefore indicates that succinate and no doubt other dehydrogenase substrates have been lost from the cells, while the dehydrogenase enzymes are still present, perhaps because of their large molecular size and the fact that some of them at least are located in the mitochondria. In much the same way failure of the tetrazolium test in tissue that has been frozen and then allowed to thaw can be attributed to a dearth of dehydrogenase substrates.[20]

The incomplete staining in imbibed embryos is thus evidence of leakage, but leaves open the question of mechanism.

The presence of membrane damaged cells. Imbibition is most rapid in the early stages so that it is the outer cotyledon tissues of an imbibing embryo that bear the full brunt of the water movement. Later, as the water front penetrates towards the cells of the interior, imbibition slows down (Fig. 1). One would therefore expect to find disrupted cells predominantly towards the outside of an imbibed embryo. The location of such cells can be established with Evans blue, a dye which is excluded from living cells with intact membranes, only entering cells in which the membranes have been disrupted. Pea embryos imbibed in 1% Evans blue began to accumulate the dye in superficial cells within a few minutes; after 2 h the dye had "permeated three to four cell layers of axes and cotyledonary tissue."[14] (In our experiments Evans blue only penetrated into the epidermis and hypodermis, amounting to less than 5% of the volume of the embryos). When intact seeds were placed in Evans blue, on the other hand, the cells of the cotyledons within accumulated little or no dye.[14]

These observations provide strong support for the proposal that membranes may be damaged in a small proportion of cells by the inrush of water.[17] The cells most at risk are those to the outside of the cotyledons--the cells first wetted during imbibition, and exposed to the fastest inflow of water. One way to test this proposition is to reduce the initial rate of imbibition--which should do less harm to the membranes and so reduce the extent of leakage.

Retarded imbibition. The presence of the testa slows imbibition in peas (Fig. 1). Evans blue is excluded from cotyledon cells and the initial rate of leakage is reduced. These observations are consistent with the proposal that the slow entry of water does no damage to cell membranes--but it should be noted that with slower imbibition fewer cells will be wetted in unit time, which will itself reduce the rate of leakage. The seeds of peanut (*Arachis hypogea* L.) behave differently; although the thin testa has no effects on the rate of imbibition, it nevertheless causes a marked reduction in leakage rate,[13] and the exclusion of Evans blue from cotyledon cells. It must be concluded that the effects of the testa on leakage are complex, and not to be understood solely on a basis of effects on imbibition rate.

A second way of slowing imbibition is by reducing the potential gradient between embryo and water. This can be achieved by first allowing embryos to become slightly wet, by exposing them to moist air or damp filter paper. If such embryos are then immersed in water, the rate of imbibition is reduced accordingly (Fig. 2). Embryos placed directly in water imbibe 0.62 g water g^{-1} embryo dry wt in an hour, and leakage in that period raises the conductivity of the water to 460 μmho. By comparison, embryos which have been wetted for 12 h and are then transferred to water imbibe only 0.44 g g^{-1} in the next hour and lose only 100 μmho. This reduction in leakage rate could be ascribed to the reduced rate of imbibition, but other explanations are possible: the outer cells which are wetted when a dry embryo is placed in water may be more susceptible to damage than those reached by the water front as it penetrates to the interior and/or the translocation of solutes leaked out of the superficial cells may well proceed with greater efficiency than the movement of solutes from inner cells. In short, these results (and comparable experiments with soybean[19]) do not on their own offer unequivocal support for the view that the initial phase of leakage results from membrane damage associated with rapid imbibition. However they do emphasize that leakage can continue (albeit somewhat slowly) under conditions of slow imbibition and at a time when the wetting front has passed beyond the outermost cells.

Another way of slowing imbibition is to place dry embryos in solutions of low water potential or high viscosity such as 10 M lithium chloride or 55% sucrose. During the first hour imbibition in 55% sucrose is reduced by 45% as compared to the controls while the leakage of potassium is six times less (Fig. 3). This reduced leakage could be ascribed to the slower imbibition, but again other possibilities arise: as imbibition is slow in sucrose fewer cells are wetted in the first hour than in water controls, and if each cell makes the same contribution to leakage, the total amount of leakage must be less than in water; in addition the solution may interfere with the process of translocation of solutes from the innermost cells. What is most striking about these experiments, however, is that eventually at the end of the 300 h period the total amount of leakage is as great in the sucrose solution as in water. We have obtained comparable results with experiments in which pea embryos were set to imbibe in 10 M lithium chloride or 6 M calcium chloride. The lithium chloride solution slowed the rate of leakage of sugars and UV-absorbing solutes as well as potassium, but the final amounts lost from the embryos after 300 h was not reduced.

Experiments of this kind, designed to reduce the rate of imbibition could clearly be repeated with a whole variety of solutes. Powell and Matthews[18] have used polyethylene

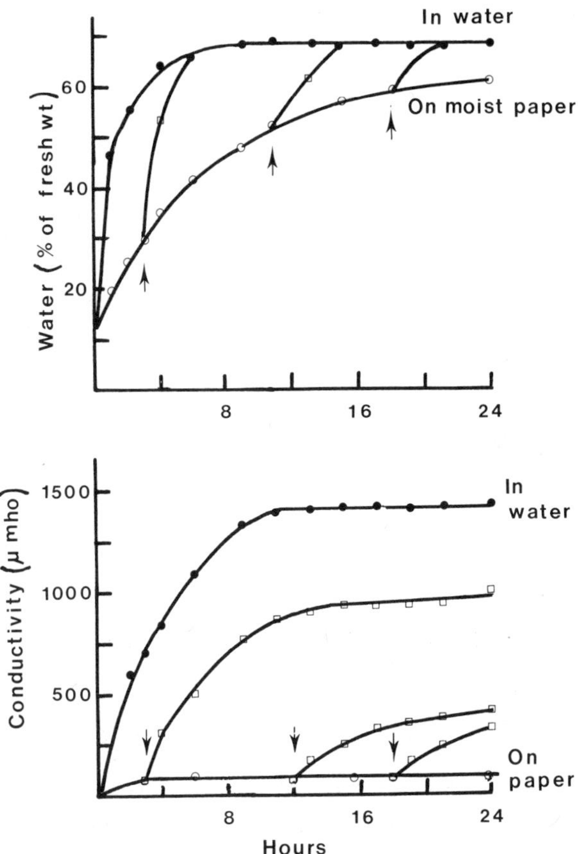

Fig. 2. Time course of imbibition and electrolyte leakage from pea embryos. —●— 50 embryos placed in 100 ml water from the start of the experiment. —○— Batches of 50 embryos imbibed on moist filter paper for the time shown and then given a 2-min wash to remove surface solutes; the washings were made up to 100 ml and their conductivity added to the conductivity of the filter paper also immersed in 100 ml water. The arrows indicate times at which batches of 50 embryos initially imbibing on filter paper were transferred to 100 ml water. 1981 seed.

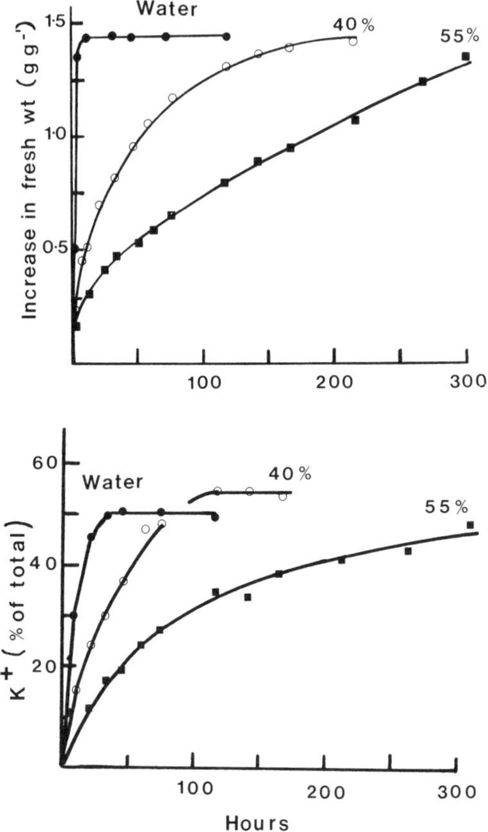

Fig. 3. Time course of imbibition and leakage of potassium from pea embryos. One batch of 50 embryos was imbibed in 100 ml water until 125 h by which time the water had become contaminated. Batches of embryos were also set up to imbibe in sucrose at 40 and 55% (wt sucrose/wt solution). 1981 seed.

glycol (PEG) which forms viscous solutions in water and is an effective osmotic agent with intact pea seeds.[21] A 30% solution of PEG of molecular weight 4000 which has a water potential of -19 bars reduced the rate of imbibition by peas in the first hour by about 20%. Embryos imbibed in this solution for 24 h gave a more complete tetrazolium reaction than those imbibed in water, indicating a reduced outflow of dehydrogenase substrates. Experiments in which embryos were immersed in water for periods up to 30 min and then transferred to 30% PEG indicated that "damage" (as revealed by a poor reaction to tetrazolium) could be incurred after no more than 2 min exposure to water. This reinforces the view that membrane rupture is only likely to occur in the outermost cells of the cotyledons, for they are reached while the inrush of water is at its most vigorous.

Conclusions. The Evans blue experiments clearly support the hypothesis that in peas the inrush of water causes irreversible damage to cells around the periphery of imbibing cotyledons. These cells receive the full force of the incoming water and they are damaged in the first minutes of imbibition. Treatments aimed at moderating the rate of imbibition result in slower leakage in part because fewer cells are wetted over the initial period, and in part because the membranes in the peripheral cells can withstand the more gentle flow of water.

However, there remain a number of observations which are not readily understood on this basis. Damage is restricted to the peripheral cells which occupy no more than 5% of the cotyledon volume. Why then does leakage continue for a matter of hours, long after the outer cells have been wetted? How is it that embryos may lose fully 60% of their potassium especially in view of the fact that cobaltinitrite staining indicates that potassium is more or less evenly distributed within the cotyledons? Why is there a loss of dehydrogenase substrates from the outer 40 to 50% of the volume of the cotyledons as evidenced by the tetrazolium test? And finally, if damage results from the sheer force of the incoming water, why is there eventually as much leakage even when the rate of imbibition has been reduced by a substantial fraction (Fig. 3)?

On the grounds that leakage continues for long periods, involves more than superficial cells and continues even when imbibition is quite slow, it seems that some other, additional mechanism must be at work.

Membrane Reorganization

The problems raised above can be accommodated on the proposition that membranes are disorganized in dry seeds, no longer forming an intact barrier around the cytoplasm of each cell--but regaining their normal semi-permeable condition during imbibition. It is envisaged that there would be a short period at the start of imbibition when the membrane constituents in each cell were going through a phase of reorganization and solutes could leak out. During the course of imbibition the water front penetrates slowly into the body of a seed or embryo, wetting each layer of cells in turn, and so allowing leakage from cell after cell. Leakage would therefore continue over a prolonged period and would embrace many cells, not just those on the periphery. The essence of this view is that the plasma membrane of dry seeds differs in some way from that found in hydrated seeds.

Our present view of membrane structure is basically that of Singer and Nicolson,[22] a lipid bilayer with intrinsic proteins floating in it, either located to one side of the membrane or spanning across the membrane from side to side. The molecular orientation of the phospholipids and proteins in the membrane is determined by their amphipathic nature. The polar and charged regions are oriented towards the aqueous phases that lie on either side of the membrane, while the central domain is a hydrophobic one inhabited by the hydrocarbon tails of the lipids and the essentially non-polar regions of the proteins. This model is widely accepted as a basis for the molecular architecture of membranes in plant and animal cells. These cells have ample water to ordain and stabilize the bilayer configuration. But what is the situation in air-dry seeds, characterized by a very low water poential and the absence of any continuous aqueous phase? Is the bilayer arrangement still a realistic guide to the arrangement of the phospholipid molecules under these conditions?

Some clues can be derived from biophysical studies, in particular thermal analysis and X-ray diffraction. Thermal analysis makes it possible to detect the presence of free water in a preparation by the exotherm that is formed at 0°C as the water freezes. Chapman[23] found that free water was present in phospholipid/water mixtures containing ample water, but not in mixtures which had less than 20% water. Under these conditions all the water was bound to the phospholipids, none remaining free to freeze at 0°C. What happens when the water content falls below this critical amount can be judged from X-ray analysis. Luzzati and Husson[24] have described the liquid-crystalline phases formed by a phospholipid extract from human brain. At high water contents there was an ordered sequence of bilayers separated by aqueous phases. As the water content of the mixture was reduced the aqueous phases became narrower, but below 20% water a completely different arrangement was found, the hexagonal mesophase. This is a predominantly hydrophobic phase, pierced by long water-filled channels disposed in a hexagonal arrangement and lined throughout by the polar heads of the phospholipids.

The concept of the hexagonal mesophase formed the basis of a hypothesis[4,12] about the structural changes that membranes were thought to undergo as they developed and finally dried out, and the reverse change that was presumed to happen during imbibition. In essence, the membrane of the developing seed corresponds to one of the bilayers seen in bulk phospholipid/water mixtures (Fig. 4, left). As the seed matures, it dries out. At the time when its water content falls below about 20% the molecular architecture of the membrane constituents is thought to change, forming a small portion of the hexagonal phase (Fig. 4, right). When this idea was first put forward[4] the hexagonal phase was described as "porous", a term which could be taken to imply that the watery channels had a transverse orientation, allowing free access for instance from the cytoplasm to the exterior of the cell. This interpretation is incorrect as McKersie and Stinson[11] rightly point out. The water-filled channels run along the plane of the 'membrane', not across it. It would be meaningless to speak of the permeability of the membrane in this dry, hexagonal condition because the

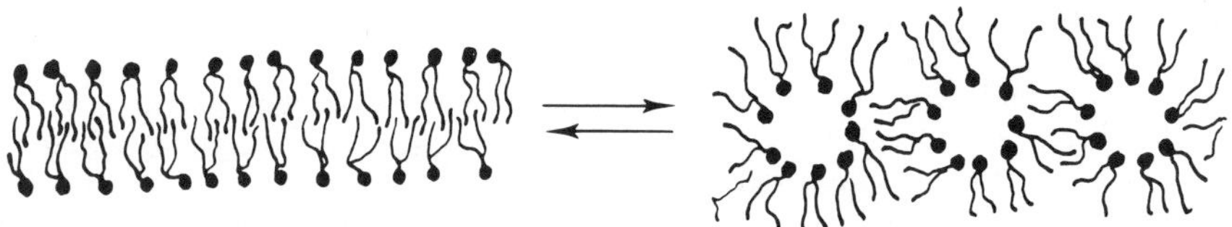

Fig. 4. Left: Simplified model of membrane structure which ignores the presence of proteins. The polar head groups of the phospholipids face the aqueous phase on either side of the bilayer. Right: hypothetical model of the conformation adopted by the phospholipids in dry seeds. The small amount of water present is located in long channels running at right angles to the plane of the paper, and lined by the head groups of the phospholipids.

addition of water to test permeability would be sufficient to convert the membrane back to the bilayer condition. This is indeed what is presumed to happen on imbibition. It is the upheaval occasioned by the reorientation of phospholipid molecules as they progress from the arrangement shown on the right in Fig. 4 to that on the left that allows the loss of solutes. The period of permeability in any one cell is normally short and comes to a halt as soon as the bilayer arrangement is restored.

If the inflow of water during imbibition is particularly rapid (as in the outer cells of an embryo) then some phospholipid molecules might be swept so far from their original location that the membrane would never become re-established, remaining in a fragmented state. The hypothesis could thus encompass that of irreversible membrane rupture and account for the entry of Evans blue into the outer cells of imbibing embryos.

The hypothesis is thus in line with many of the observations about leakage from seeds and embryos; it can also provide a basis for understanding the hydrodynamics of pollen.[25,26] The model shown in Fig. 4 can be extended by considering the location of the intrinsic membrane proteins as well as the phospholipids. These proteins are normally located in the membrane by virtue of their amphipathic nature. However as a seed becomes dehydrated they are likely to follow the phospholipids, breaking away from their normal location as their polar heads seek to face such relatively aqueous regions as persist in the 'dry' seed. Upon imbibition these proteins will ideally become located back in the membrane, in their original positions. It has been argued elsewhere[12,12a] that this incorporation of proteins back into the membrane corresponds to the 'maturation' of mitochondria in imbibing peas, a process in which the protein complement of mitochondria rises by the incorporation of pre-existing cytoplasmic proteins. (Peanuts are unlike peas in this respect, for their mitochondria do not have to undergo such a maturation during imbibition.[27] The inference would be that their intrinsic proteins remain in place, even in the dry seeds).

As a general concept, the hypothesis can thus claim to account for a number of the phenomena observed in imbibing peas and pollen. In more detail, at the level of molecular biology, the scheme of Fig. 4 is open to criticism; it is surely over-simplified and may be quite wrong. There are, first of all unlikely to be any long water-filled channels in a dry seed in view of its very low water potential and the profusion of strongly hydrophilic molecules present. As the seed matures and dries out the phospholipids will be in an environment of falling water potential and increasing viscosity. Moreover the environment will not be homogenous, but one that varies considerably from one micro-locality to another in keeping with the great structural complexity of cytoplasm itself. Perhaps it is sufficient to suggest that as the seed dries out some regions will remain relatively moist for longer than others. If they are close enough, such moist spots will attract the polar head groups of phospholipids. Fig. 4 suggests that the arrangement of phospholipids is regular and orderly in both hydrated and dry seeds; in reality, the arrangement in dry seeds is likely to be far less regular and orderly than the diagram suggests. (An additional factor governing lipid distribution may be the expansion that the plasma membrane undergoes as a seed swells during imbibition; a recent report[28] suggests that this expansion occurs by incorporation of lipids and proteins lying nearby, that were presumably displaced from the membrane as the seed contracted in size during its final maturation.)

A more damaging criticism stems from the fact that the original concept is based on work with brain phospholipid preparations.[24] Attempts to detect the presence of a hexagonal phase in phospholipid preparations from seeds of *Lotus*[11] and soybean[29] have proved unsuccessful. If the seed phospholipids do form a hexagonal phase it is evidently rather elusive; the same conclusion emerges from a ^{31}P-NMR study of pollen.[29a] It may be that the ratio of phosphatidylcholine to phosphatidylethanolamine is so much higher in the preparations from these particular seeds than in brain phospholipids as to militate against the hexagonal conformation.[30]

If membrane constituents were to adopt the hexagonal phase over more or less extensive areas under dry conditions, one might wonder whether this could be detected by electron microscopy. It is unlikely that the standard techniques using aqueous fixatives would be a reliable guide to cell structure in dry seeds for the tissue would probably become more or less hydrated *before* it was fixed. Unfortunately the results obtained with non-aqueous fixatives are somewhat equivocal, Opik[31] reporting that the plasma membrane was continuous and unbroken in dry rice fixed with gaseous osmium tetroxide, while Thomson[32] observed "ill-defined" membranes in dry mung bean seeds fixed in formaldehyde-glycerol. In freeze-fracture studies Thomson and Platt-Aloia[32a] find that the plasma membrane in dry cowpea seeds is highly convoluted but reveals a fairly normal fracture plane, suggesting that "the membrane is a bilayer and leakiness may be more related to the degree of order within the bilayered membrane than to a primary restructuring and/or reorganization of the membrane components in the dry state." The difficulty of assessing the molecular arrangement of phospholipids and proteins from an observed electron microscope image is indeed a recurring theme in this field. However Buttrose[33] has compared the appearance under freeze-fracture of pieces of barley scutellum frozen in the dry state or after 8 sec exposure to water. The dry tissue looks so different in these preparations from the hydrated, as to encourage the view that "membrane" architecture is quite different in the dry and hydrated states.

The same view emerges from a recent study of soybeans.[28] Tissue with less than 18% water yielded only small areas identifiable as plasma membrane and these contained many irregularities or pock marks; as imbibition proceeded larger expanses of plasma membrane were revealed, with fewer irregularities and more particles embedded in the membrane sheets. In addition the appearance of lettuce seed in freeze-fracture preparations depends on water content,[34] in such a way as to suggest a membrane phase change at a water content above 20 to 25%.

CONCLUSIONS

We propose, then, that membrane structure is disorganized in dry tissues, perhaps in an irregular fashion, and perhaps only in certain regions. The peripheral cells of isolated embryos are subject to such a violent inrush of water during imbibition that functional membranes never reappear and there is no barrier to the entry of Evans blue or the loss of solutes. The cells of intact seeds on the other hand, and the inner cells of isolated embryos imbibe water more slowly, and in them membranes become re-established once more--but only after a short intervening period of phospholipid reorientation during which leakage is possible.

These ideas relate to the movement of solutes out of cells. In the case of spores and perhaps pollen this may be sufficient to account for the appearance of solutes in the bathing medium. However in a bulky seed a further consideration comes into play--the mechanism involved in transporting solutes released by an inner cell out to the bathing medium. How is it that solutes can move outwards against the inflow of water? It has been commonly supposed that diffusion is the mechanism in question.[4,12] We are currently investigating the basis for this view.

ACKNOWLEDGMENT

One of us (LKM) wishes to thank the Department of Education for Northern Ireland for financial support.

REFERENCES

1. SHAYKEWICH CF, J WILLIAMS 1971 Resistance to water absorption in germinating rapeseed (Brassica napus L). J Exp Bot 22: 19-24
2. HARPER JL, RA BENTON 1966 The behaviour of seeds in soil. II. The germination of seeds on the surface of a water supplying substrate. J Ecol 54: 151-166
3. WAGGONER PE, J-Y PARLANGE 1976 Water uptake and water diffusivity of seeds. Plant Physiol 57: 153-156
4. SIMON EW 1974 Phospholipids and plant membrane permeability. New Phytol 73: 377-420
5. POWELL AA, S MATTHEWS 1979 The influence of testa condition on the imbibition and vigour of seeds. J Exp Bot 30: 193-197
6. MATTHEWS S, AA POWELL, NE ROGERSON 1980 Physiological aspects of the development and storage of pea seeds and their significance to seed production. In PD Hebblethwaite, ed, Seed Production, Butterworths, London pp 513-525
7. SIMON EW, RM RAJA HARUN 1972 Leakage during seed imbibition. J Exp Bot 23: 1076-1085
8. MATTHEWS S, NE ROGERSON 1976 The influence of embryo condition on the leaching of solutes from pea seeds. J Exp Bot 27: 961-968
9. POWELL AA, S MATTHEWS 1981 A physical explanation for solute leakage from dry pea embryos during imbibition. J. Exp Bot 32: 1045-1050
10. BRAMLAGE WJ, AC LEOPOLD, DJ PARRISH 1978 Chilling stress to soybeans during imbibition. Plant Physiol 61: 525-529
11. McKERSIE BD, RH STINSON 1980 Effect of dehydration on leakage and membrane structure in Lotus corniculatus L seeds. Plant Physiol 66: 316-320

12. SIMON EW 1978 Membranes in dry and imbibing seeds. In JH Crowe, JS Clegg, eds, Dry Biological Systems, Academic Press, New York pp 205-224
12a. SIMON EW 1983 Respiration and membrane reorganization during imbibition. In JM Palmer, ed, The Physiology and Biochemistry of Plant Respiration, Soc Exper Biol Seminar Ser, Cambridge Univ Press in press
13. ABDEL SAMAD IM, RS PEARCE 1978 Leaching of ions, organic molecules, and enzymes from seeds of peanut (*Arachis hypogea* L) imbibing without testas or with intact testas. J Exp Bot 29: 1471-1478.
14. DUKE SH, G KAKEFUDA 1981 Role of the testa in preventing cellular rupture during imbibition of legume seeds. Plant Physiol 67: 449-456
15. DUNN BL, RL OBENDORF, DJ PAOLILLO 1980 Imbibitional surface damage in isolated hypocotyl-root axes of soybean (*Glycine max* (L) Merr cv chippewa 64). Plant Physiol 65: S-139
16. SIMON EW, HH WIEBE 1975 Leakage during imbibition, resistance to damage at low temperature and the water content of peas. New Phytol 74: 407-411
17. LARSON LA 1968 The effect soaking pea seeds with or without seed coats has on seedling growth. Plant Physiol 43: 255-259
18. POWELL AA, S MATTHEWS 1978 The damaging effect of water on dry pea embryos during imbibition. J. Exp Bot 29: 1215-1229
19. TULLY RE, ME MUSGRAVE, AC LEOPOLD 1981 The seed coat as a control of imbibitional chilling injury. Crop Sci 21: 312-317
20. STEPONKUS PL 1971 Effect of freezing on dehydrogenase activity and reduction of triphenyl tetrazolium chloride. Cryobiology 8: 570-573
21. MANOHAR MS 1966 Effect of "osmotic" systems on germination of peas (*Pisum sativum* L). Planta 71: 81-86
22. SINGER SJ, GL NICOLSON 1972 The fluid mosaic model of the structure of cell membranes. Science 175: 720-731
23. CHAPMAN D, DFH WALLACH 1968 Recent physical studies of phospholipids and natural membranes. In D Chapman, ed, Biological Membranes, Physical Fact and Function, Academic Press, London pp 125-202
24. LUZZATI V, F HUSSON 1962 The structure of the liquid-crystalline phases of lipid-water systems. J Cell Biol 12: 207-219
25. HESLOP-HARRISON J 1979 An interpretation of the hydrodynamics of pollen. Am J Bot 66: 737-743

26. SHIVANNA KR, J HESLOP-HARRISON 1981 Membrane state and pollen viability. Ann Bot 47: 759-770

27. MOROHASHI Y, JD BEWLEY, EC YEUNG 1981 Biogenesis of mitochondria in imbibed peanut cotyledons: influence of the axis. J Exp Bot 32: 605-613

28. CHABOT JF, AC LEOPOLD 1982 Ultrastructural changes of membranes with hydration in soybean seeds. Am J Bot 69: 623-633

29. SEEWALT V, DA PRIESTLEY, AC LEOPOLD, GW FEIGENSON, F GOODSAID-ZALDUONDO 1981 Membrane organization in soybean seeds during hydration. Planta 152: 19-23

29a. PRIESTLEY DA, B KRUIJFF 1982 Phospholipid motional characteristics in a dry biological system. A ^{31}P-nuclear magnetic resonance study of hydrating _Typha ratifolia_ pollen. Plant Physiol 70: 1075-1078

30. HUI SW, TP STEWART, PL YEAGLE, AD ALBERT 1981 Bilayer to non-bilayer transition in mixtures of phosphatidylethanolamine and phosphatidylcholine: implications for membrane properties. Arch Biochem Biophys 207: 227-240

31. OPIK H 1980 The ultrastructure of coleoptile cells in dry rice (_Oryza sativa_ L) grains after anhydrous fixation with osmium tetroxide vapour. New Phytol 85: 521-529

32. THOMSON WW 1979 Ultrastructure of dry seed tissue after a non-aqueous primary fixation. New Phytol 82: 207-212

32a. THOMSON WW, K PLATT-ALOIA 1982 Ultrastructure and membrane permeability in cowpea seeds. Plant Cell Environ 5: 367-373

33. BUTTROSE MS 1973 Rapid water uptake and structural changes in imbibing seed tissues. Protoplasma 77: 111-122

34. TOIVIO-KINNUCAN MA, C STUSHNOFF 1981 Lipid participation in intracellular freezing avoidance mechanisms of lettuce seeds. Cryobiology 18: 72-78

Chapter Three

MEMBRANE STRUCTURE IN GERMINATING SEEDS

B. D. McKERSIE AND T. SENARATNA

Department of Crop Science
University of Guelph
Guelph, Ontario
Canada N1G 2W1

INTRODUCTION

The structure and function of cellular membranes in seeds are dynamic. Substantive changes occur as water is imbibed, as the axis initiates elongation, and, at least in dicot seeds, as the reserves are mobilized from the cotyledons. This brief summary seeks to review some of the more recent experiments concerning the structure of the cellular membranes in seeds during each of these processes.

CHANGES IN MEMBRANE STRUCTURE DURING IMBIBITION

The hydration of a dry seed is accompanied by a large efflux of cytoplasmic solutes including inorganic ions, amino acids, sugars, phosphate, proteins, and other cytoplasmic solutes.[1-3] The quantity of these solutes, especially electrolytes, has been negatively correlated in many studies with seed viability and seedling vigour[2-5] and thus has implications from a practical point of view in commercial seed testing.[6]

Kinetic analysis of solute efflux has demonstrated that the time profile for the appearance of solutes in the imbibing solution is biphasic (Fig. 1). Immediately after the transfer of a dry seed into an imbibing solution, there is a rapid efflux of solutes from the seed. Within minutes, the rate of solute leakage diminishes and is gradually replaced by a much slower but constant rate of solute leakage.[2,3] Several explanations have been proposed to account for this biphasic pattern; for example, changes in membrane properties during hydration and the physical rupture of cells caused by the rapid inrush of water have been outlined in the previous chapter. Perhaps the most controversial is that of Simon[2] who proposed that at moisture contents below 20%, there is not sufficient water present to maintain the typical lamellar or bilayer structure and that as a consequence, the phospholipids are rearranged into a hexagonal phase (Fig. 2). The experimental evidence suggesting this model was a low angle x-ray diffraction study by Luzzatti and Husson[7] which examined the effects of hydration on the phase properties of phospholipid extracted from mammalian tissues. A characteristic of the hexagonal phase proposed by their study was the presence of long-water filled channels (Fig. 2).

Simon[2] recognized that if such a phase occurs in dry seed membranes, it could explain the rapid efflux of solutes during the first minutes of imbibition. With increasing water

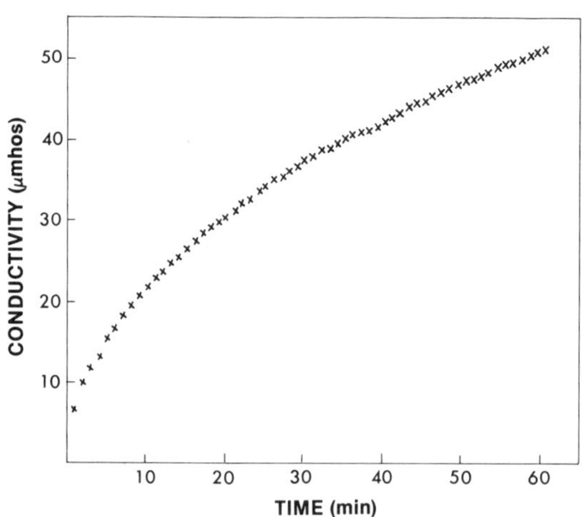

Fig. 1. Time profile of electrolyte leakage from _Lotus corniculatus_ seeds during imbibition. Adapted from reference 12.

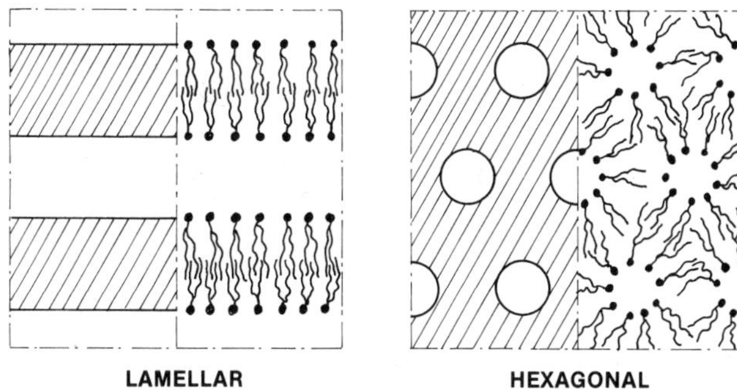

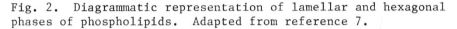

Fig. 2. Diagrammatic representation of lamellar and hexagonal phases of phospholipids. Adapted from reference 7.

content and reformation of the lamellar phase, these channels would disappear and the rate of solute leakage from the cytoplasm would decrease. The assumptions inherent in Simon's model are: 1) that the component phospholipids of the cellular membranes in mammalian tissues and in seeds respond to hydration in a similar manner and 2) that the biphasic pattern of leakage from imbibing seeds is dictated by rapid changes in membrane permeability. Recently, these two assumptions have been tested experimentally and have been shown to be invalid.

To test the first assumption, we repeated Luzzatti and Husson's earlier experiments but using phospholipids extracted from dry seeds.[1] _Lotus corniculatus_ seeds were ground into a fine powder, extracted with organic solvents and the lipid extract was partitioned into polar and neutral lipid fractions. The neutral lipid fraction was discarded because of the probability that it contained significant quantities of triglycerides. The polar lipid extract was dried _in vacuo_, weighed and hydrated to 5, 10, 20 and 40% water content. The lipid-water system was incubated at room temperature under a nitrogen atmosphere for 24 h and then transferred to a sealed chamber on the low angle x-ray diffraction camera.

Lamellar packing can be distinguished from hexagonal packing on the basis of the ratio between the first, second, third, and fourth order Bragg spacings.[7] In a lamellar phase, the ratios are 1:1/2:1/3:1/4, whereas in a hexagonal phase the ratios are $1:1/\sqrt{3}:1/\sqrt{4}:1/\sqrt{7}$. All phospholipid-water preparations analyzed had ratios characteristic of a lamellar structure. Even at 5% hydration, the ratios of the Bragg spacings were those of a lamellar phase with no evidence for the presence of hexagonally packed phospholipids at any moisture content examined. Since the original experiments of Luzzati and Husson were performed with lipid extracts from human brain, it must be assumed that the observed differences in response to hydration by the two systems were a result of differences in lipid composition.

Confirmation of this initial observation has come recently using three quite different techniques. Firstly, ^{31}P-nuclear magnetic resonance measurements of the phospholipids in dry *Typha latifolia* (cattail) pollen could not detect a hexagonal phase at moisture levels as low as 11%.[8] Secondly, electron micrographs of the radicle from dry cotton seeds which were fixed without hydration indicated the presence of a continuous, intact plasmalemma.[9] Thirdly, freeze fracture studies of the same tissue yielded micrographs typical of membranes in a lamellar phase.[9] Thus, it would appear that plant phospholipids respond to low water contents quite differently from mammalian phospholipids, and typically do not form a hexagonal phase.

The second assumption inherent in Simon's model, namely that the biphasic pattern of electrolyte leakage is a consequence of changes in membrane permeability, has also been questioned. Powell and Matthews[10] killed pea embryos by heating at 105°C for 24 h and observed that the rate of electrolyte leakage consistently followed a biphasic pattern through 4 consecutive imbibe/dry cycles (Fig. 3). Since the membranes in these heat-treated embryos should be readily permeable to electrolytes, the authors concluded that the biphasic pattern of electrolyte leakage must be attributed to the changing physical nature of the imbibing seed and not to changes in the permeability of cellular membranes.

To understand how the physical nature of the seed generates this biphasic pattern, consider a model system consisting of a spherical seed immersed in a large volume of water. The initial hydration of a seed proceeds as a hydration front moving gradually from the outside to the inside of the seed.[11] Thus, at any given time prior to full hydration, dry and hydrated regions exist within the seed (Fig. 4) and hence the hydrated volume (V_h), or the number of hydrated cells in the partially imbibed seed, can be estimated as the difference in volume of two spheres:

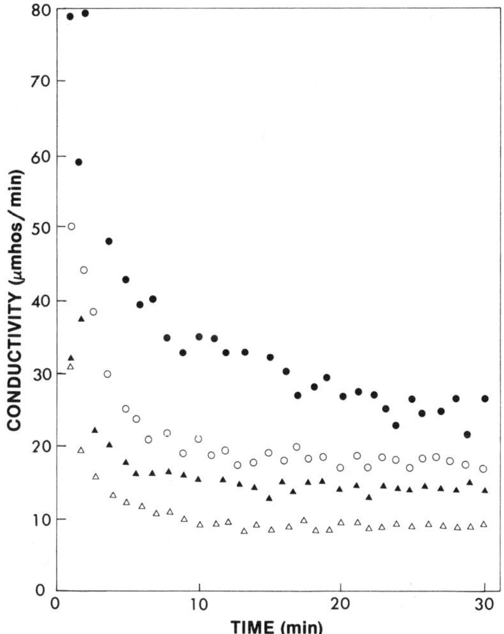

Fig. 3. Flux of electrolytes from dead pea embryos following four imbibe/dry cycles: (●) cycle 1; (○) cycle 2; (▲) cycle 3; (△) cycle 4. Adapted from reference 10.

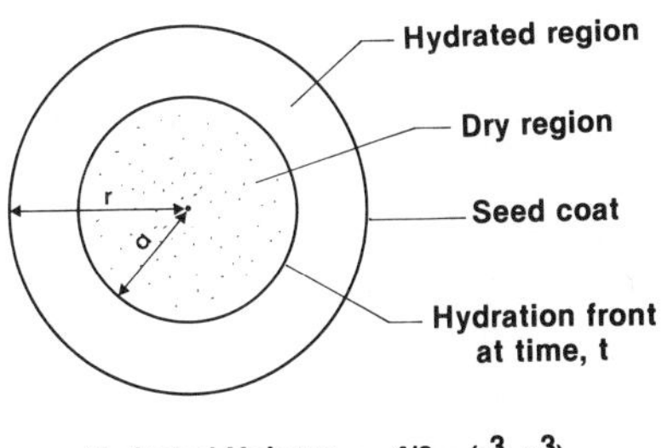

Fig. 4. Schematic diagram of a model spherical seed showing hydrated and dry regions at time, t, shortly after imbibition. Adapted from the data and figures in reference 11.

$$V_h = \frac{4}{3}\Pi(r^3 - a^3)$$

where r is the radius of the seed, and a is the radius of the dry region of the seed. If the hydration front advances at a constant rate, a decreases linearly, and V_h increases exponentially (Fig. 5). Leakage of solutes from this model will occur from the hydrated region of the seed, the solutes in the dry region being immobile in the absence of solvent. Therefore, with time the quantity of solutes able to leak from the seed increases in proportion to V_h.

Within the hydrated region of the seed, solutes leak from two sites--the free space outside of a diffusion-limiting membrane, and the cytoplasm inside a diffusion-limiting

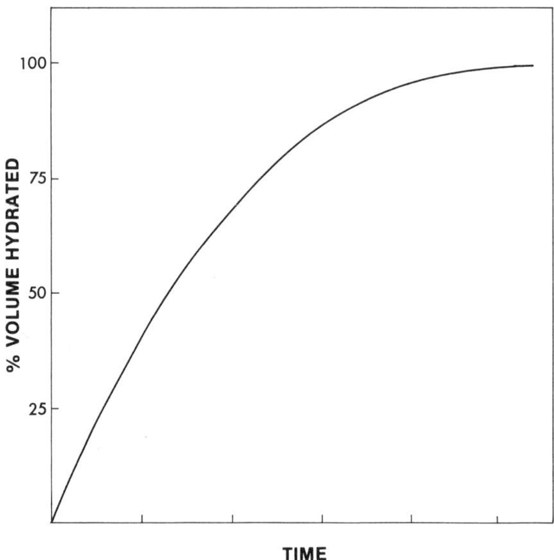

Fig. 5. Hydrated volume (%) of a model spherical seed at various times after transfer into water. Calculated from the relationships

$$V_h = \frac{4}{3}\Pi(r^3 - a^3) \text{ and } a = r - k \cdot t,$$

where k is the rate of water penetration into the seed. See text for explanation.

membrane. The rate of leakage, expressed in units such as moles seed^{-1} min^{-1}, can thus be expressed as:

$$J_T = J_a + J_c \qquad (1)$$

where J_T is the net flux into the imbibing solution, J_a is the flux of solutes originating from the free space and J_c is the flux originating from the cytoplasm. J_a is defined by the law of diffusion as:

$$J_a = -D\cdot\left(\frac{dc}{dx}\right)_a\cdot V_h \qquad (2)$$

where D is the diffusion constant and $(dc/dx)_a$ is the concentration gradient from the inside to the outside of the seed. Since D is comparatively large for most solutes, dc/dx will decrease rapidly. Therefore, J_a will be limited by the availability of solutes in the aqueous solution, which is dependent on the hydrated volume at any given time. Thus, at a first approximation, the rate of leakage from the free space is a function of the rate of hydration:

$$J_a = f\left(\frac{dV_h}{dt}\right) \qquad (3)$$

The rate of leakage from the cytoplasmic compartment, J_c, is defined as:

$$J_c = -P\cdot\left(\frac{dc}{dx}\right)_c\cdot A \qquad (4)$$

where P is the permeability coefficient of the diffusion-limiting membrane, $(dc/dx)_c$ is the concentration gradient across the membrane and A is an estimate of the total cell surface area in the seed which is hydrated. Therefore, A is a function of V_h. Assuming that P is small relative to D, the rate of leakage from the cytoplasmic compartment will continue for a significant period of time. If we also assume that P and $(dc/dx)_c$ are similar for all cells in the seed, then at a first approximation, the rate of leakage from the cytoplasmic compartment is a function of the hydrated volume:

$$J_c = f(V_h) \qquad (5)$$

J_c will become constant once the seed is hydrated, and will continue until there no longer is a concentration gradient.

If equations (1), (3) and (5) are combined, then:

$$J_T = f\left(\frac{dV_h}{dt}\right) + f(V_h)$$

Thus, at a first approximation, the rate of leakage from an imbibing seed which is measured experimentally (J_T) is the sum of two functions: one dependent on the rate of hydration and the second on the hydrated volume at any given time. These relationships are shown graphically in Fig. 6. With

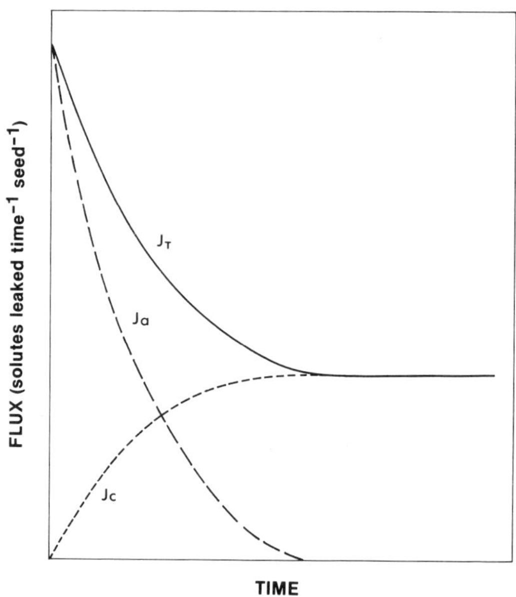

Fig. 6. Flux of solutes from free space (J_a) and cytoplasmic (J_c) compartments during imbibition based on the relationships

$$J_a = f\left(\frac{dV_h}{dt}\right)$$

and $J_c = f(V_h)$ where V_h is an estimate of hydrated cell number. The net flux J_T is the sum of J_a and J_c.

time the solute flux originating in the free space (J_a) decreases to zero as these spaces are hydrated and leached of solutes. Concurrently the flux from the cytoplasm (J_c) increases in proportion to the number of hydrated cells. The sum of these two components approximates a theoretical net flux (J_T) expressed as solute leaked time^{-1} seed^{-1}, which is very similar to the experimental profile observed for an imbibing seed (Fig. 3). Integration of this curve gives a typical biphasic pattern (Fig. 7). In this model, as observed experimentally,[10] the biphasic pattern of leakage is not dependent on changes in membrane permeability during hydration, but instead is related to differences in degree of hydration, rates of diffusion and solute concentration gradients.

Thus, at present there is no experimental evidence to support the concept that membrane lipids form a hexagonal phase in the dry seed. Although it seems logical to propose that membranes, which maintain their structure as a result of the hydrophobic-hydrophilic interactions among phospholipids,

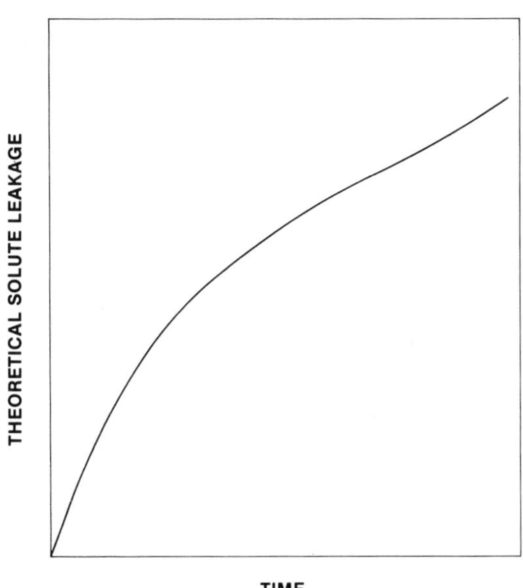

Fig. 7. Theoretical time profile of solute leakage from a model spherical seed assuming the relationships in Fig. 6. Line represents the integration of net flux in Fig. 6.

would be altered at low water contents, the physical nature of these changes at low water contents remain to be identified.

CHANGES IN MEMBRANE STRUCTURE DURING AXIS ELONGATION

After the seed is hydrated, many metabolic events are initiated in both the axis and cotyledon which convert the dicot seed from a dormant or non-growing embryo into an actively growing seedling. Coincident with the initiation of growth in the axis is a loss of desiccation tolerance. At physiological maturity, seeds can tolerate desiccation to low moisture levels, but after a critical stage in germination, they lose their tolerance and become susceptible to dehydration injury.[1,12] For example, *Lotus corniculatus* seeds can be imbibed for 6 h to full hydration, dried to 10% moisture and reimbibed without impairing germination or seedling vigour (Fig. 8). However, a similar treatment imposed on seeds which had been imbibed for 24 h completely prevented subsequent germination. This transition to desiccation sensitivity is coincident with radicle emergence in both *Lotus corniculatus* and *Glycine max*. In these species, injury occurred only when the seeds were dried below 20% moisture (Fig. 9), but in other more sensitive species injury has been recently reported to occur at moisture levels as high as 55%.[13]

In addition to preventing germination, desiccation of sensitive seeds disrupts membrane integrity as indicated by increased leakage of many cytoplasmic solutes (Fig. 8 and Fig. 9). This has led to our hypothesis that the primary effect of desiccation is at the level of the membrane and that this injury to the membrane systems of the seed impairs germination. Furthermore, we have hypothesized that the loss of desiccation tolerance in the seed is a consequence of changes in membrane properties which occur during germination. However, the nature of injury to the membranes and the nature of the changes which accompany the development of sensitivity are as yet unknown. The following text is a brief outline of some experiments designed to test our hypotheses and to describe more precisely the changes which occur in membrane properties after desiccation.

Although the transition to desiccation sensitivity is coincident with radicle emergence, the loss of desiccation tolerance can occur in the absence of cell enlargement. For

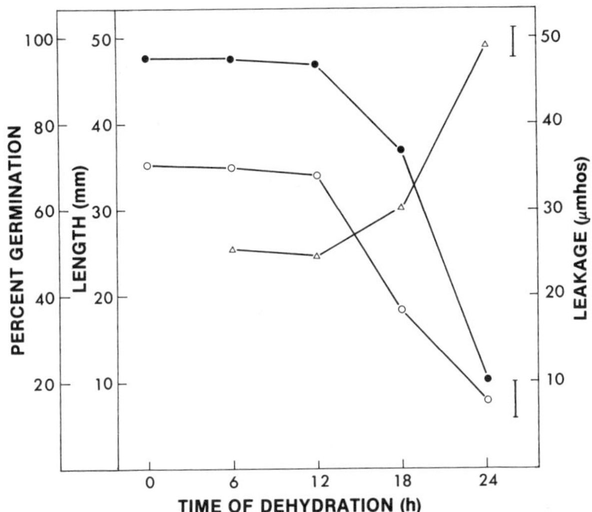

Fig. 8. Response of Lotus corniculatus L. (cv. Carroll) seeds to dehydration. Seeds were incubated for 0, 6, 12, 18, and 24 h, dehydrated, and reimbibed. Percent germination (●) and total seedling length (O) were determined after 5 days. Leakage (Δ) represents the conductivity (micromhos) of 10 ml imbibing solution after imbibition of 50 dehydrated seeds for 60 min. Values represent the means of four determinations. Least significant differences at the 5% level are shown as vertical bars for length and leakage measurements. Data from reference 12.

example, axes of G. max seeds imbibed in polyethylene glycol (PEG) or cycloheximide solution have not elongated by 36 h imbibition, but if these seeds are dehydrated, they are no longer viable and the axes exhibit high rates of electrolyte leakage (Table 1). Consequently, preventing cell elongation with PEG or cycloheximide did not prevent the loss of desiccation tolerance. The results of the kinetic analysis of electrolyte leakage gives the typical biphasic pattern of leakage from the dehydrated axes which can be characterized by two components: the y-intercept and the slope of the linear portion of the biphasic profile. The former estimates the quantity of solutes leaked from extracellular sites such as the seed surface, intracellular spaces or ruptured cells, whereas the latter estimates the rate of flux of solutes across a diffusion-limiting membrane.[1,14] Axes dehydrated

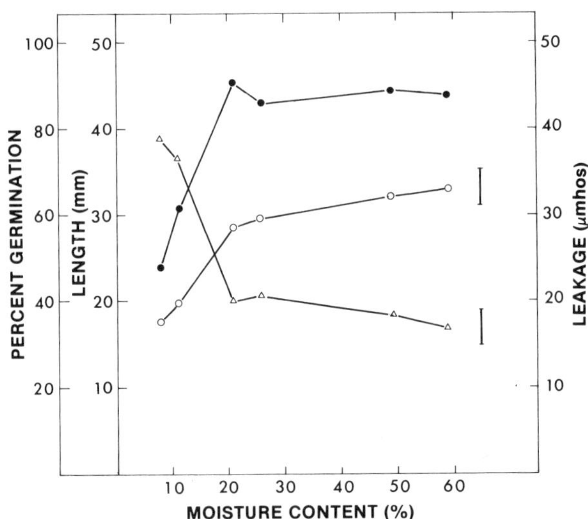

Fig. 9. Response of Lotus corniculatus L. (cv. Carroll) seeds to varying degrees of dehydration. Seeds were imbibed for 24 h, dehydrated to specific moisture contents, and reimbibed. Percent germination (●) and total seedling length (O) were determined after 5 days. Leakage (Δ) represents the conductivity (micromhos) of 10 ml imbibing solution after imbibition of 50 dehydrated seeds for 60 min. Values represent the means of three or four determinations. Least significant differences at the 5% level are shown as vertical bars for length and leakage measurements. Data from reference 12.

after the initiation of elongation (36 h) exhibited higher initial leakage rates (y-intercept) and higher rates of leakage compared to axes dehydrated at 6 h (Table 1). This implies that there are two components to desiccation injury--an increase in cell rupture, and an increase in membrane permeability. When elongation was prevented with PEG or cycloheximide, the axes dehydrated at 36 h exhibited less cellular rupture during rehydration, but similar changes in membrane permeability. Since the axes dehydrated after PEG or cycloheximide treatment were no longer viable, the increase in membrane permeability would appear to be more critical for the survival of dehydrated axes than the increased incidence of cell rupture.

Table 1. Effect of desiccation on seed germination and leakage from axes of *Glycine max* after imbibition for 6 or 36 h in water, polyethylene glycol (PEG) or cycloheximide solutions.

Seeds were imbibed for 6 h in distilled water, for 36 h in distilled water, for 6 h in distilled water followed by 30 h in PEG at -6 bars, or for 36 h in 20 μg/ml cycloheximide. All seeds were subsequently dehydrated to 8% moisture. Percent germination was determined after 5 days rehydration of intact seeds. Conductivity of the imbibing solution was measured at 15, 30, 45, 60, 120, 180, 240, 300, 360 and 420 min soaking periods. The y-intercept and slope of the linear regression line was calculated from electrolyte leakage values between 2-7 h soaking periods.

Time of dehydration	Imbibing solution	Axis elongation	Germination	Initial leakage (y-intercept)*	Rate of leakage (slope)*
h			%	(μmhos·100 mg^{-1})	(μmhos·100 mg^{-1})
6	water	absent	93 a	27 c	10 b
36	water	present	0 b	323 a	27 a
36	PEG	absent	0 b	84 b	26 a
36	CHI	absent	ND	57 b	25 a

* values within a column followed by the same letter are not significantly different at $p \leq 0.05$

ND - not determined

Further evidence that membrane permeability has been altered in desiccation injured axes is shown by the following three experiments.

1) Measurements of the rate of efflux of various solutes from axes which were desiccated and subsequently fully rehydrated indicate that different solutes leak at different rates, and that this difference is maintained, but quantitatively altered, in desiccation-injured tissue (Table 2).[15] When expressed relative to the total solute concentration, K^+ leaks at a faster rate than the other solutes. In desiccation-injured tissue, the rate of K^+, sugar and protein efflux increased 2-fold, amino acid 30-fold and phosphate leakage 100-fold. This differential increase in leakage of various solutes supports the concept that the majority of these solutes are moving across a semi-permeable membrane and that the permeability of these membranes has been altered by desiccation.

2) Measurements of the Arrhenius activation energy (E_a) for solute efflux also provide a measure of membrane permeability.[15] A high E_a value indicates a greater permeability barrier than a low E_a. Thus the lower E_a values for the efflux of K^+, phosphate and amino acid from dehydration injured tissue indicate that its permeability barrier was reduced (Table 3). However, the E_a for protein was unchanged implying that either the efflux of protein was not limited by membrane permeability or that the membrane had remained intact and continued to limit protein efflux.

3) Measurements of the effects of extracellular pH on K^+ and sugar efflux have also been performed.[15] In axes dehydrated at 36 h imbibition, and rehydrated, a reduction in extracellular pH promoted the efflux rate of K^+ but decreased the efflux rate of sugar (Fig. 10).

Collectively, these experiments imply that the plasmalemma, or whichever membrane system is limiting solute leakage, remains intact in the majority of cells in the axis after lethal desiccation injury. Although membrane permeability has apparently been altered, some functions remain such as sensitivity to pH gradient. We are presently attempting to determine if the injury which occurs to cellular membranes involves perturbation of the lipid bilayer or disruption of the transport proteins in the membranes. In addition we are attempting to define the changes in these membranes which render them sensitive to desiccation injury.

CHANGES IN MEMBRANE STRUCTURE DURING MOBILIZATION OF STORAGE RESERVES

The cotyledons of dicotyledonous seeds serve as the major storage organ for the heterotrophic seedling. For example, in etiolated Phaseolus vulgaris seedlings, the protein reserves are rapidly translocated out of the cotyledons between days 2 and 9 after imbibition (Fig. 11). At the same time, the storage cells undergo autolysis as indicated by a general breakdown of cytoplasmic structure. The membrane components in the cotyledons are also degraded and the total phospholipid (PL) which could be extracted from the smooth microsomal fraction from the cotyledons declines (Fig. 11).

The functional properties of the smooth microsomes change as indicated by component enzymatic activities and permeability properties. The activities of microsomal NADH cytochrome c reductase and glucose-6-phosphatase decline

Table 2. Rate of solute efflux from fully hydrated Glycine max axes.

Axes were imbibed for 6 or 36 h, and dehydrated to 8% moisture. Ten dehydrated axes were reimbibed in 10 ml distilled water and the quantity of solutes in the imbibing solution was quantified at time intervals between 2 and 8 h soaking periods. Values were expressed relative to total solute concentration in solution after 10 axes were homogenized in 10 ml distilled water.

Solute	Solute leaked	
	6 h	36 h
	$\% \, h^{-1}$	
Potassium	6.0	15.0
Phosphate	0.03	3.00
Amino acid	0.1	3.4
Sugar	3.3	5.8
Protein	0.5	1.0

All values are significantly different ($p \leq 0.05$) between 6 and 36 h axes according to analysis of variance.

Table 3. Arrhenius activation energies for efflux of solutes from fully hydrated *Glycine max* axes.

Axes were imbibed for 6 or 36 h, dehydrated to 8% moisture and rehydrated for 2 h in distilled water. Ten axes were then transferred to 10 ml distilled water preincubated at 6 temperatures between 10 and 35°C. The solution was decanted after 4 h incubation and analyzed for solute concentration. Arrhenius activation energy was calculated as:

$$E_a = 2.303\,R \frac{\log K_2 - \log K_1}{1/T_1 - 1/T_2}$$

where K_1 and K_2 are rates of solute leakage at the absolute temperatures T_1 and T_2, respectively. R is the gas constant.

	6 h	36 h
	Kcal mol^{-1}	
Potassium	11.3	5.7*
Phosphate	11.0	8.4*
Amino acid	7.4	4.3*
Protein	4.9	5.3 NS

*, NS - significantly different between 6 and 36 h treatments at $p \leq 0.05$, and not significantly different, respectively, according to analysis of variance.

asynchronously as the reserves are depleted (Fig. 12). Barber and Thompson[18] have shown that the permeability of liposomes prepared from the lipid extracted from the smooth microsomal fraction increases between days 2 and 7 (Fig. 13).

The explanation for some of the functional changes may lie in observations made on the fluidity and phase properties of these microsomal membranes.[19-22] Electron spin resonance measurements using the probe 2N14 indicate that their viscosity increases as the tissue senesces (Fig. 14). Furthermore, the liquid-crystalline to gel phase transition temperature detected by wide angle x-ray diffraction increases from a low of approximately 2°C at day 2 to over 50°C at day 9 (Fig. 15). Consequently, at the growth temperature of 29°C, increasing proportions of the membrane lipid are in the gel phase. By analogy to other experimental systems, this change in the fluidity and phase properties could explain the increased membrane permeability and decreased enzymatic activity.[21,22]

In an attempt to determine the compositional basis for the alteration in membrane viscosity and phase properties, the lipid was extracted from isolated microsomal membranes and used to reconstitute artificial membranes or liposomes. The transition temperatures of liposomes made from the total lipid extract were only slightly below that of the intact membrane (Fig. 16). However, if the liposomes were reconstituted from the phospholipid fraction, the large increase in transition temperature was not observed. This implies that the neutral lipids which were removed from the total lipid extract during purification of the phospholipid fraction determined the phase transition temperature of the phospho-

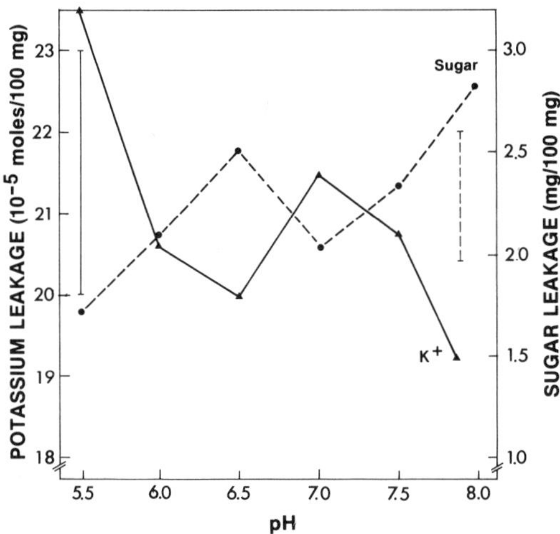

Fig. 10. Effect of extracellular pH on potassium (▲) and sugar (●) leakage. Glycine max axes were imbibed for 36 h and dried to 8% moisture. Axes were reimbibed for 2 h, and 10 axes were transferred to 25 ml MES-HEPES buffer at appropriate pH. Solute concentration in the buffer was measured between 1 and 4 h and rate of efflux calculated. Least significant differences at the 5% level are shown by vertical bars. Data from reference 15.

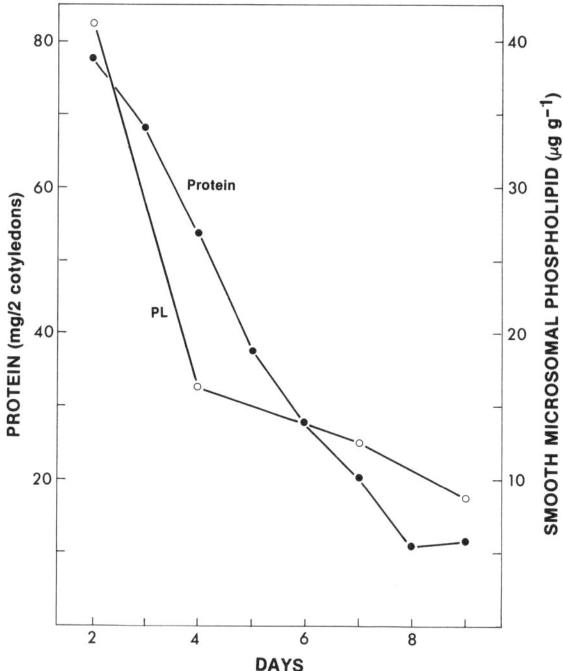

Fig. 11. Total protein (●) and smooth microsomal phospholipid (○) content of Phaseolus vulgaris cotyledons during growth at 29°C in the dark. Protein data adapted from reference 16 and phospholipid data from reference 17.

lipids. Confirmation of this was obtained by mixing phospholipids extracted from a 2 day old microsomal fraction with neutral lipid from 2 or 9 day microsomes. As the concentration of neutral lipid in the liposome increased so did the transition temperature (Fig. 16). The neutral lipid from the 9 day old tissue was more effective than that from the 2 day old tissue on a percent weight basis suggesting compositional differences. Presumably, a component in the neutral lipid induced a phase separation of the phospholipids creating gel and liquid-crystalline domains within the membrane.

The physiological implications of these changes are not clear. It is possible that the increased viscosity and incidence of gel phase lipid represent stages in membrane disassembly. It is, however, equally possible that these

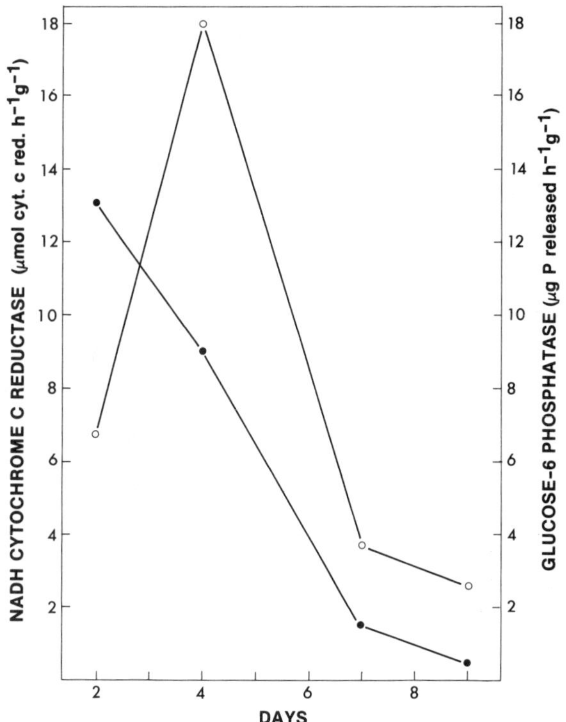

Fig. 12. NADH cytochrome c reductase (●) and glucose-6-phosphatase (○) activities in the smooth microsomal fraction from Phaseolus vulgaris cotyledons during growth at 29°C in the dark. Data adapted from reference 9.

changes are metabolically programmed to induce functional changes in the membrane such as decreased enzymatic activity, increased permeability or the release of compartmentalized hydrolytic enzymes which mediate the hydrolysis of storage reserves and the eventual autolysis of the storage cell.

SUMMARY

Some of the changes in membrane structure and function in germinating dicotyledonous seeds have been discussed. No experimental evidence is available to support the hypothesis that there is a hexagonal-lamellar phase transition during seed imbibition. Instead the membrane lipids of viable dry seeds appear to be packed in a lamellar phase. If present in seeds, the hexagonal phase must be restricted to membrane systems with peculiar lipid compositions. The biphasic pattern of leakage, which suggested that changes in membrane permeability occur during seed imbibition, can be related mathematically to the rate of hydration and the hydrated seed volume.

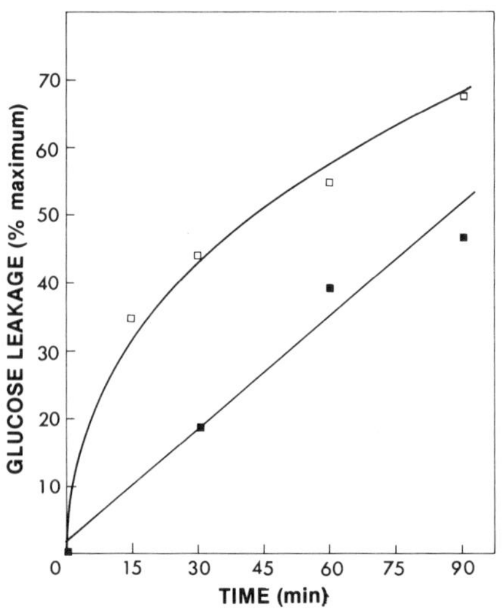

Fig. 13. Profiles of glucose leakage from liposomes prepared from the total lipid extracts of smooth microsomal membranes of Phaseolus vulgaris cotyledons isolated at 2 days (■) and 7 days (□). Data from reference 8.

Fig. 14. Viscosity of smooth microsomal membranes from Phaseolus vulgaris cotyledons measured with the ESR probe 2N14. Adapted from reference 21.

Fig. 15. Liquid-crystalline to gel phase transition temperatures of smooth microsomal membranes of Phaseolus vulgaris cotyledons (○), liposomes prepared from the total lipid extract of the smooth microsomes (●) and liposomes prepared from the phospholipid extract (▲) as detected by wide angle x-ray diffraction. Adapted from references 20 and 22.

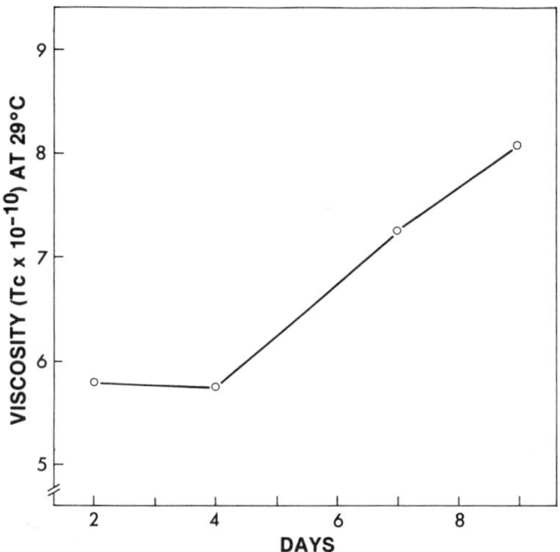

Fig. 14 (see legend on facing page)

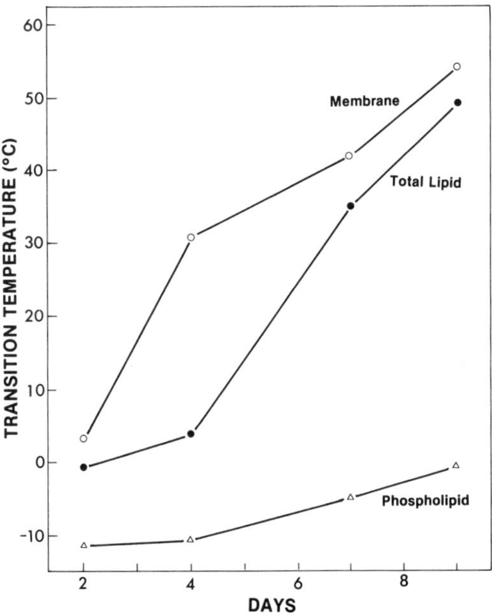

Fig. 15 (see legend on facing page)

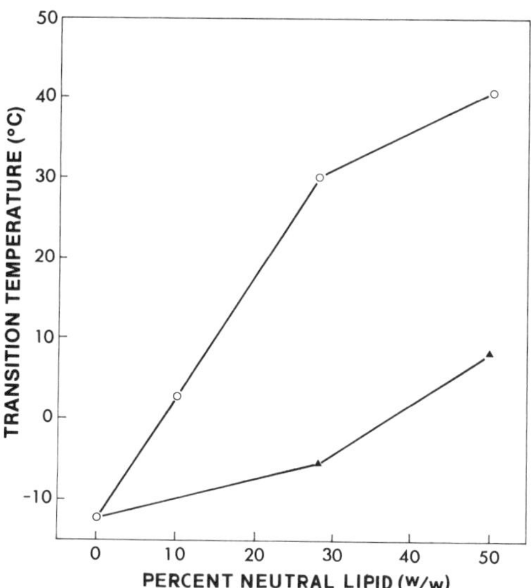

Fig. 16. Effect of neutral lipid fraction isolated from smooth microsomes from 2 day-old (▲) and 9 day-old (O) Phaseolus vulgaris cotyledons on the liquid-crystalline to gel phase transition temperature of liposomes prepared from phospholipid from 2 day-old, smooth microsomes. Adapted from reference 22.

Prior to elongation, the axis loses its tolerance of desiccation and this may be related to physical or compositional changes of cellular membranes or to a protective system, but is not related to changes of cell size. Desiccation injury in sensitive seeds is characterized by an increase in membrane permeability but it remains to be established if this is due to a disruption of the lipid bilayer or to an inactivation of the ion pumps.

During mobilization of the reserves from cotyledons, increasing proportions of the membrane lipid are present in a gel phase. Rearrangement of the phospholipid in the membrane bilayer to cause a phase separation is mediated by increasing proportions of neutral lipid in the membrane. The incidence of gel phase lipid may be related to increased membrane permeability, release of compartmentalized hydrolytic enzymes, and inactivation of membrane bound enzymes.

REFERENCES

1. McKERSIE BD, RH STINSON 1980 Effect of dehydration treatment on leakage and membrane structure in Lotus corniculatus L. seeds. Plant Physiol 66: 316-320
2. SIMON EW 1974 Phospholipids and plant membrane permeability. New Phytol 73: 377-420
3. SIMON EW, RM RAJA HARUN 1972 Leakage during seed imbibition. J Exp Bot 23: 1076-1085
4. BRAMLAGE WJ, AC LEOPOLD, DJ PARISH 1978 Chilling stress to soybeans during imbibition. Plant Physiol 61: 525-529
5. McKERSIE BD, DT TOMES, S YAMAMOTO 1981 Effect of seed size on germination, seedling vigour, electrolyte leakage and establishment of birdsfoot trefoil (Lotus corniculatus L.). Can J Plant Sci 61: 337-343
6. MATTHEW S 1981 Evaluation of techniques for germination and vigour studies. Seed Sci Technol 9: 543-551
7. LUZZATI V, F HUSSON 1962 The structure of the liquid crystalline phases of lipid-water system. J Cell Biol 12: 207-219
8. PRIESTLEY DA, B de KRUIJFF 1982 Phospholipid motional characteristics in Typha pollen determined non-invasively by ^{31}P-nuclear magnetic resonance. Plant Physiol 69(Suppl): 47
9. VIGIL EL, MN CHRISTIANSON, RL STEERE, WP WERGEIN 1982 Dynamic endomembrane complex in radicle cells of cotton seeds. Plant Physiol 69(Suppl): 3
10. POWELL AA, S MATTHEWS 1981 A physical explanation for solute leakage from dry pea embryos during imbibition. J Exp Bot 32: 1045-1050
11. WAGGNER PE, JY PARLANGE 1976 Water uptake and water diffusivity of seeds. Plant Physiol 57: 153-156
12. McKERSIE BD, DT TOMES 1980 Effect of dehydration treatments on germination seedling vigour and cytoplasmic leakage in wild oats and birdsfoot trefoil. Can J Bot 58: 471-476
13. BECWAR MR, PC STANWOOD, EE ROSS 1982 Dehydration effects on imbibitional leakage from desiccation sensitive seeds. Plant Physiol 69: 1132-1135
14. SENARATNA T, BD McKERSIE Dehydration injury in germinating soybean (Glycine max. (L.) Merr.) seeds. Plant Physiol (manuscript submitted)
15. SENARATNA T, BD McKERSIE Characterization of solute efflux from dehydration injured soybean (Glycine max.

(L.) Merr.) seeds. Plant Physiol (manuscript submitted)

16. LEES GL, JE THOMPSON 1975 The effect of germination on the subcellular distribution of cholinesterase in cotyledons of Phaseolus vulgaris. Physiol Plant 34: 230-237
17. McKERSIE BD, JE THOMPSON 1975 Cytoplasmic membrane senescence in bean cotyledons. Phytochemistry 14: 1485-1491
18. BARBER RF, JE THOMPSON 1980 Senescence-dependent increase in the permeability of liposomes prepared from bean cotyledons. J Exp Bot 31: 1305-1313
19. McKERSIE BD, JE THOMPSON, JK BRANDON 1976 X-ray diffraction evidence for decreased lipid fluidity in senescent membranes from cotyledons. Can J Bot 54: 1074-1078
20. McKERSIE BD, JE THOMPSON 1977 Lipid crystallization in senescent membranes from cotyledons. Plant Physiol 59: 803-807
21. McKERSIE BD, JR LEPOCK, J KRUUV, JE THOMPSON 1978 The effects of cotyledon senescence on the composition and physical properties of membrane lipid. Biochim Biophys Acta 508: 197-212
22. McKERSIE BD, JE THOMPSON 1979 Phase properties of senescing plant membranes. Role of the neutral lipids. Biochim Biophys Acta 550: 48-58

Chapter Four

REGULATION OF NITROGEN METABOLISM DURING EARLY SEEDLING GROWTH

ANN OAKS

Biology Department
McMaster University
Hamilton, Ontario
Canada L8S 4K1

INTRODUCTION

During grain development the basic storage reserves, (carbohydrates, lipids, and proteins) are laid down in all parts of the grain or seed. After germination hydrolysis of these reserves is seen first in the seedling, subsequently in the cotyledon or scutellum and finally, in the endosperm.[1,2] The carbohydrate, whether stored initially or derived from lipids,[3] is available as a substrate to supply the energy and reducing power as well as the building blocks required for the formation of new cells (Fig. 1). One place where metabolism involved in the reworking of carbon is potentially regulated is in the reutilization of the amino acids supplied by the hydrolysis of storage proteins.

Seed proteins are characterized by an unbalanced distribution of amino acids. For example, legume seeds are deficient in sulfur amino acids but are rich in arginine; cereal seeds are deficient in tryptophan and in the basic amino acids, principally lysine, but are rich in proline, leucine, glutamine and asparagine. Thus some amino acids are supplied in amounts sufficient or almost sufficient to meet the demands

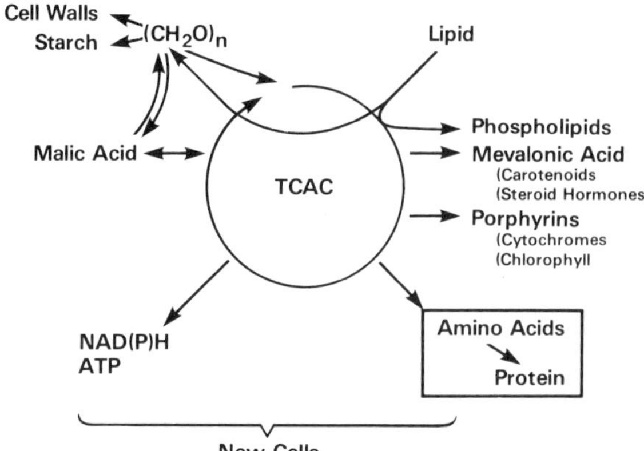

Fig. 1. The utilization of starch and lipid as carbon and energy reserve in the seedling system. (CH_2O) represents carbohydrates in general, TCAC, the tricarboxylic acid cycle.

imposed by the synthesis of seedling proteins while others are supplied in excess of those demands. Folkes and Yemm[4] examined the amino acid balance in barley seedlings during the initial ten days of growth (Table 1). They found that glutamate, proline and asparagine disappeared during the growth period presumably to supply carbon and nitrogen for the synthesis of required amino acids and other nitrogen containing compounds. This has been referred to in the literature as the interconversion of amino acids. Most of the other amino acids showed only minor losses or gains. Subsequently Joy and Folkes[5] using cultured barley embryos and Oaks[6,7] using maize root tips showed that the synthesis of many amino acids (pro, arg, thr, lys, ileu, val, leu) was regulated specifically by the abundance of the end product. Sodek and Wilson[8,9] using a maize mutant rich in lysine (opaque-2) and its wild type sibling were able to show that the degree of endogenous synthesis during seed development or germination was moderated by the amino acid supply. For example, they found that embryo leucine was derived entirely from the endosperm whereas lysine was synthesized largely within the embryo.[9] Although many of the amino acids supplied by the endosperm are utilized by the embryo, none of the amino acids is essential for seedling growth as indicated by the relatively simple media required to support embryo cultures.[10-12]

Table 1. Change in total amino acid content during germination in barley.[a]

Amino Acid	Day 0	Change in 10 days
Glutamic family		
Glu	11.37[b]	-6.42
Pro	7.22	-4.58
Arg	7.71	+1.05
Aspartic family		
Asn	10.87	-6.52
Asp	3.13	+1.20
Thr	1.79	+0.51
Lys	3.01	+2.04
Met	0.66	+0.04
Branch chain amino acids		
iLeu	2.48	+0.12
Val	2.97	+0.01
Leu	3.52	-0.06
Aromatic amino acids		
Phe	2.14	-0.36
Tyr	1.26	+0.06
Trp	0.67	+0.41

[a] Adapted from reference 4.
[b] Values are expressed as mg/seed (endosperm plus embryo) at day 0 and as the change in mg/seedling (endosperm plus seedling) at day 10.

In this chapter I will concentrate on cereals, principally on *Zea mays*. However, I would like to stress that observations we have made with *Zea mays* are probably not unique to *Zea mays* or indeed to cereals. I will consider four aspects of the controls which are important during that phase of seedling development when endosperm nitrogen reserves are still contributing to the nitrogen economy of the seedling. 1. Influence of carbohydrates on utilization of nitrogen reserves; 2. Regulation of the hydrolysis of the storage proteins; 3. Regulation of the biosynthesis of amino acids; 4. Regulation of the assimilation of nitrate.

INFLUENCE OF CARBOHYDRATES ON UTILIZATION OF NITROGEN RESERVES

Initially hydrolysis and utilization of storage reserves occurs in the same cells. However, by the time endosperm reserves are required, transport of hydrolysis products is probably rate limiting. It seems likely, therefore, that the rate of supply of glucose or sucrose could limit growth and development of the young seedling. For example, induction of nitrate reductase, an energy requiring process, is enhanced in root tips by the addition of glucose.[13-15] In excised root tips where we know that the endogenous level of glucose is rapidly depleted[16] additions of glucose have an even more dramatic effect on the induction of nitrate reductase.[17]

The potential complexity of the effect of glucose on nitrogen metabolism is summarized in Table 2. There is a marked drop in both glucose/sucrose levels and in proline content in root tips during the 6 h period after excision. Glucose additions to the culture medium reduce both the absolute loss of proline and the oxidation of [UL-^{14}C]proline. Thus the oxidation of alternative substrates, potentially involved in the production of ATP and NAD(P)H, is influenced by glucose levels in the cell. Glucose additions also lead to increases in protein synthesis as measured by the incorpor-

Table 2. The effect of glucose additions on the aspects of nitrogen metabolism in maize root tips.

	Intact Root Tip		Excised Root Tip	
	- glucose units[1]	+ glucose % change[2]	- glucose units[1]	+ glucose % change[2]
Glucose (μmoles/40 tips)	20.8	-	7.8	-
Proline (μmoles/40 tips)	1.7	-	0.38	132
Protein synthesis (cpm×10^{-3})	500	-	9800	140
Oxidized proline (cpm×10^{-3})	-	-	686	64
Nitrate Reductase	155	294	55.5	375
Glutamate dehydrogenase	1091	-	2603	60
Asparagine synthetase	4.2	-	15.6	57

[1]Roots were incubated for 6 hr. in presence or absence of glucose (1%) in 0.1 strength Hoaglands salts which contained 10 mM KNO_3. Total protein synthesis is the incorporation of label from [2 ^{14}C]-acetate into amino acids of the alcohol insoluble residue during a 2 hr. incubation period. Oxidized proline is the sum of [^{14}C]-proline converted to organic acids, glutamate, aspartate and their amides during a 30 min. incubation with [UL-^{14}C]-proline. Nitrate reductase, glutamate dehydrogenase and asparagine synthetase levels are in nmoles of substrate used or product formed per mg protein in 1 hr. (Compiled from references 7,13,16,17).

[2]% change considers values in - glucose treatment (control) as 100; in the + glucose treatment as a % of that control value.

ation of [^{14}C]acetate into the alcohol insoluble residue. However individual proteins are not equally affected. For example, nitrate reductase activity, an indicator of anabolic reactions in the cell, is enhanced 7-fold by the addition of glucose, whereas glutamate dehydrogenase or asparagine synthetase activities, potential indicators of rapid cell turnover and senescence, are repressed.

HYDROLYSIS OF THE STORAGE PROTEINS

Proteins represent 10 to 16% of the dry weight of cereal endosperm. The prolamine fraction (zein, in maize) which is rich in proline, leucine, glutamine and asparagine and the glutelin fraction which has a more balanced amino acid composition, are the major proteins in the cereal endosperm,[18] and they serve as the main source of reduced nitrogen available to the young seedling.[4,9,10,19,20] A combination of endo-, carboxy- and amino-peptidase activities,[21-23] are involved in the hydrolysis of these proteins. Appearance of endopeptidase activity parallels the loss of endosperm proteins in both intact grains and detached endosperm pieces[19] suggesting that activity is important in reserve protein breakdown. The endopeptidase(s) can hydrolyze a wide range of protein substrates including urea-denatured zein and gliadin.[24] In addition, a peptide hydrolase with the properties of this endopeptidase has been purified from corn endosperm.[25] The appearance of carboxypeptidase activity coincides with the appearance of endopeptidase activity.[21,22] These two types of activity are apparently responsible for the major hydrolysis of the storage proteins in maize.

The results in Table 3 summarize the effects of the hormones gibberellic acid (GA_3) and abscisic acid (ABA), and inhibitors of protein (cycloheximide) and RNA (cordycepin) synthesis on the induction of α-amylase and the major proteases in maize endosperm pieces. Acid endopeptidase, carboxypeptidase and α-amylase activities increase over the incubation period in the absence of any additions. As with barley, additions of cycloheximide and cordycepin inhibit the development of α-amylase. This suggests that a *de novo* synthesis of protein accompanies the appearance of enzyme activity. Cycloheximide also inhibits the appearance of the two protease activities again suggesting a *de novo* synthesis of a protein. Cordycepin on the other hand, has no effect on the development of protease activity. Since the proteases represent activities of several distinct proteins[22,23] it is

Table 3. The effect of gibberellic acid (GA_3) on the development of α-amylase and the peptide hydrolases on maize endosperm pieces.

	α-amylase[1]	peptidase hydrolases[1]				α-amylase[1]	peptidase hydrolases[1]		
		endo-	carboxy-	amino-			endo-	carboxy-	amino-
Endosperm from Mature Seed					Endosperm from Immature Seed				
Day 0	0	0.9	3.3	10.7	Day 0	0.5	3.7	4.7	47.6
Day 6	22.6(100)	6.9(100)	11.0(100)	1.9(100)	Day 4	29.7(100)	1.0(100)	3.6(100)	5.6(100)
				Values Relative to Control (%)					
GA_3 (30µM)	146	106	96	84	G_A (30µM)	211	990	72	93
ABA (2µM)	35	41	65	97					
ABA + GA_3	110	84	94	78					
Cycloheximide (18µ)	0	8	2	-					
Cordycepin (100 µM)	44	100	100	-					

[1]Endosperms from mature seeds were excised and incubated for 6 days; from seeds harvested 20 days post pollination and then air dried (immature seed), for 4 days. Units are expressed as mg substrate (α-amylase) or µg product (peptidase hydrolases) per g.fwt in 1 min.

difficult to interpret the meaning of the cordycepin effect at this time.

ABA inhibits development of the activities of the hydrolytic enzymes whereas GA_3, when added alone, has virtually no effect. When added together with ABA, it reverses the inhibition caused by ABA. When corn kernels are harvested prematurely (20 days post pollination) and air dried, they have the capacity to germinate. When endosperm pieces are taken from these samples, there is a clear GA effect. One interpretation of these results is that the relative amounts of GA and ABA are important in controlling induction of α-amylase and endopeptide- and carboxypeptide-hydrolases in *Zea mays*.[22,26,27] The hormone response is complex[28,29] (see Black et al, chapter 10, this volume) and probably involves both a genetic and a physiological component. In dormant wild oats (*Avena fatua*), for example, both embryo dormancy and the hydrolysis of endosperm reserves appear to be controlled by GA.[30] The degree of response varies from year to year. This probably reflects environmental influences on the relative levels of ABA and GA. In addition, pure lines of *Avena fatua* have been produced which are non-dormant and which can hydrolyze endosperm reserves in the absence of GA. Isogenic lines which are fully dormant and which require GA for both functions have also been isolated.[30]

Once the synthesis of the hydrolase activities (α-amylase or endopeptidase) is initiated in the maize endosperm there is apparently no real endogenous control over the total production and, in fact, the enzymes are probably over-produced. Harvey and Oaks[19] found, for example, that cycloheximide inhibits the production of either α-amylase or endopeptidase and that sufficient enzyme is produced by 50 h post-imbibition to mediate a normal hydrolysis of endosperm starch and protein (Fig. 2). In the absence of inhibitor, increases in enzyme activity are observed for at least 6 days.

Zeins, which are located in protein bodies, and glutelins, which form a structural support for protein bodies and amyloplasts, are insoluble in water.[18] One might, therefore, expect unique peptide hydrolases to initiate the hydrolysis of these two classes of proteins in much the same manner as the rather specific proteases which are active in cleaving pumpkin seed[31,32] and mung bean[33] storage protein. I suspect that the proteases that we have examined are able to digest the solubilized hydrolysis products of the initial hydrolysis and not the intact native zeins and glutelins. In a model system where dried endosperm powder was treated with commercial peptide hydrolases and sodium dodecyl sulfate (SDS), Oaks et al[26] were able to show that both pronase (an endopeptide hydrolase) and a carboxypeptidase were required for the "complete" hydrolysis of the endosperm protein. Proline was not a product of this digestion even though it represented about 11% of the amino content of the endosperm. Since it was detected after acid hydrolysis of the water extract, it must have been present in a peptide form. This result suggests that proline in the primary structure somehow limits the reaction of commercial proteases. A similar restriction of endogenous peptide hydrolases is also indicated by the absence of free proline in unhydrolyzed leachates obtained from endosperm pieces.[10]

The results in Table 4 show that the total nitrogen released from maize endosperm pieces is relatively constant over a 72 h period when the endosperms are excised 1, 2 or 3 days after imbibition. The increase in α-amino nitrogen upon hydrolysis in both the "acid" and "neutral and basic" amino acid fractions suggests that small peptides are important products of the normal hydrolysis of endosperm "storage" proteins. It is possible that a majority of these peptides are released from the endosperm and transported to the growing regions before they are completely hydrolyzed. The advantage of this could be that the amino acids moving as small peptides would be protected from oxidation while in transit. Proline, for example, is easily oxidized in many tissues,[16,35] and it would be protected from such oxidations if it were transported as a peptide. Appropriate mechanisms for the transfer of peptides from the endosperm to the scutellum are present since two groups of workers[36,37] have identified mechanisms for the uptake of peptides by cereal scutella.

Fig. 2. Effect of cycloheximide on the hydrolysis of endosperm storage products and on the development of α-amylase and acid endopeptidase activities. Solid lines are the control values (no cycloheximide); dotted lines are the values obtained after treatment with cycloheximide (5 μg/ml). Cycloheximide was added at the times indicated by the arrow. Reproduced from Harvey and Oaks[19] with permission of Plant Physiology.

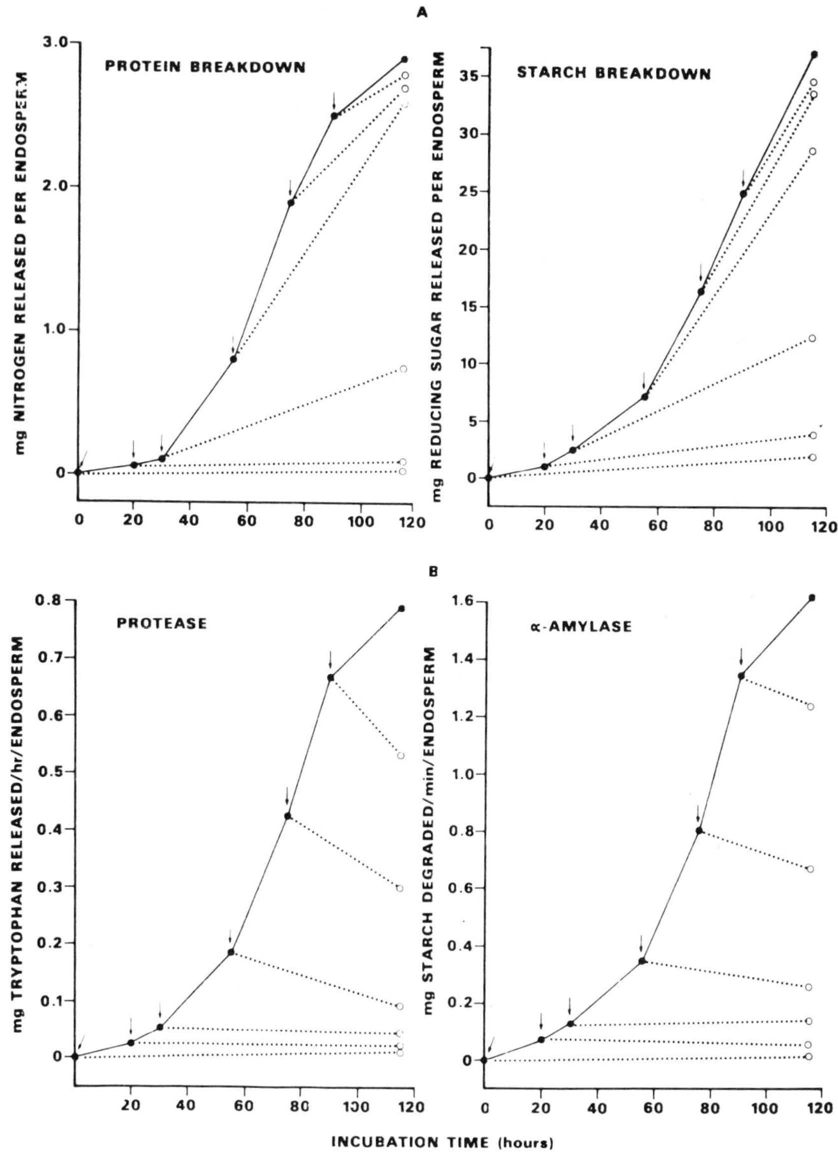

Fig. 2 (see legend on facing page)

Table 4. Evidence for peptides released from maize endosperm pieces (*Zea mays* Wf9 x 38-11).

Days after imbibition	Acidic fraction			Neutral + Basic fraction		
	Control	Acid hydrolyzed	Increase	Control	Acid hydrolyzed	Increase
	α-NH_2 N			α-NH_2 N		
	µg/10 endosperms		%	µg/10 endosperms		%
1	0.80	2.52	315	2.59	4.60	178
2	0.85	2.57	302	2.69	4.91	183
3	0.81	2.39	295	2.92	5.35	183

Endosperm pieces from 5 seeds incubated in 10 ml H_2O under sterile conditions for 72 h. The incubation medium was then filtered, hydrolyzed in 1 N HCl for 4 h at 100°C, lyophilyzed and passed over a Dowex 50 (H^+) column. The basic compounds were removed with 2 N NH_4OH after the column had been washed with water, lyophilyzed, and passed over a Dowex 1 (acetate) column. The neutral and basic peptides and amino acids were removed with an H_2O wash and the acidic amino peptides and acids with 6 N acetic acid. α-NH_2 nitrogen was measured according to the method of Yemm and Cocking[34] on the fractions as obtained from the columns (control) or after a hydrolysis in 6 N HCl at 120°C for 4 h.

REGULATION OF THE BIOSYNTHESIS OF AMINO ACIDS

An extensive analysis of the regulation of amino acid biosynthesis in seedlings or root tips has been dealt with previously.[38,39] Three recent reviews and the article by Lea and Joy (chapter 6, this volume) have also considered more general aspects of this regulation.[40-42] Basically we have found a) that the level of soluble sugars can modify nitrogen metabolism in the embryo (Table 2) and b) that certain amino acids can potentially regulate their own biosynthesis. The regulation is apparently an end product regulation of a particular enzyme activity rather than an end product repression of enzyme formation. This result has been obtained in experiments using radioactive tracers or in examining growth responses coupled with an examination of the critical enzymes.

There has been no good evidence of enzyme repression although an early observation suggested this as a reasonable explanation for the delay in protein synthesis in excised maize embryos.[15] One of the problems in identifying enzyme repression is that it has been difficult to obtain non-leaky auxotrophic mutants.[43,44] Without such mutants it is impossible to control the endogenous levels of the regulating metabolites. If there are 2 or more isoenzymes active in a particular tissue the loss or modification of one of these need not result in an auxotrophic mutant. There is evidence from Neurospora, for example, that a second isoenzyme may perform the missing function when one isoenzyme with a particular function is missing.[39] The result is a "leaky" mutation. There is now ample evidence of isoenzymes involved in amino acid biosynthetic pathways in higher plants.[42,45-48] These enzymes may differ in their primary functions[47] or in their mode of regulation.[47] Recently King et al[44] have been pursuing the time consuming task of "mutating" one isoenzyme and then "mutating" subsequent isoenzymes with the aim of obtaining a complete block at a particular biosynthetic step. Once they have been successful in obtaining mutations that block the synthesis of all of the isoenzymes, they will be in a position to examine aspects of enzyme repression with more precision.

From our experiments on the regulation of amino acid biosynthesis we know that in the case of leucine[12] and proline[38] 1) that the inhibition is specific; that is, leucine is required to inhibit leucine biosynthesis and has no effect on synthesis of other amino acids such as alanine and proline whether or not a mixture of amino acids is also present (Table 5). Proline synthesis is severely reduced by presence of the amino acid mixture, while alanine synthesis is not significantly affected (Table 5). 2) that the kinetics of inhibition are unique to each amino acid (Fig. 3). Exogenous leucine, for example, induced its maximum effect at a much lower concentration (5×10^{-5} M) than did exogenous proline (7×10^{-4} M), but was less effective in reducing the total leucine biosynthesis (to 40%) than was proline in reducing proline biosynthesis (to 10%). One fact that cannot be emphasized too strongly is that the concentrations of amino acids required to inhibit a particular reaction are in the micromolar range; that is, in a concentration range that could be expected in an <u>in vivo</u> situation. 3) that the inhibition is not the same in all tissues.[38,49] Proline, for example, effects a marked inhibition on proline synthesis in the root tip (Table 6). In more mature regions of the root exogenous proline interferes with the incorporation of endogenous proline into protein (insoluble proline) but not appreciably with the total synthesis (soluble plus insoluble proline) of proline. Other parameters (concentrations exert-

Table 5. The effect of exogenous leucine on the endogenous synthesis of leucine, proline, and alanine in maize root tips.

Leucine concentration	Sucrose (1%)			Sucrose (1%) plus corn mixture of amino acids		
	Leu	Pro	Ala	Leu	Pro	Ala
μM	cpm/40 tips			cpm/40 tips		
0	6891	2475	4405	5410	324	3680
12	3491	2485	4351	5062	427	3411
96	2520	2315	4456	1395	382	3860

Excised root tips prepared under sterile conditions were pretreated for 3 h in a 0.1 strength Hoagland's salts solution which contained 1% sucrose and the appropriate amino acids. Similar concentrations of leucine were added to each treatment. At this time the roots were washed, transferred to fresh media which contained the same ingredients plus [2-^{14}C]acetate and incubated for an additional 2 h. (Adapted from reference 7.)

ing an initial effect, half maximal and maximal effect) are basically similar in the two types of tissue. Our interpretation of this result is that there is an isoenzyme involved in

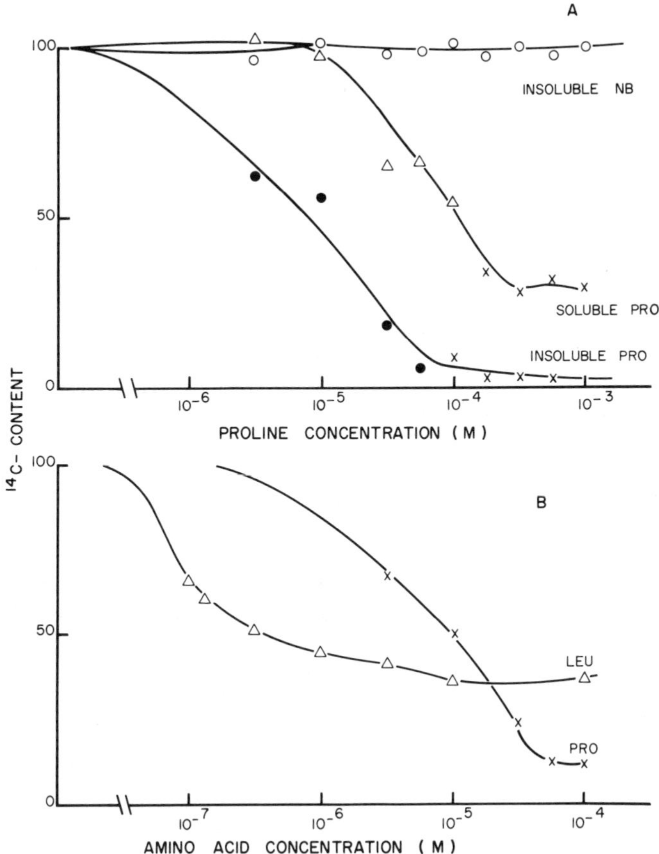

Fig. 3. Effect of exogenous leucine or proline on the endogenous biosynthesis of leucine and proline. Reproduced from Oaks et al[38] with permission of the Canadian Journal of Botany. Protocol as described in Table 5. A: alcohol soluble and insoluble [^{14}C]proline are considered separately; NB, the total radioactivity in the neutral and basic amino acid fraction is not influenced by the proline additions. B: total ^{14}C (alcohol soluble and insoluble fraction) in leucine (Δ) or proline (**X**). Values obtained with no exogenous amino acid were set at 100%, and other values are relative to these.

Table 6. The change in the capacity of proline to inhibit its own biosynthesis as root cells age.

	Root sections[a]			
	Tips (0-5 mm)		Mature sections (20-25 mm)	
	-proline	+proline (10^{-3} M)	-proline	+proline (10^{-3} M)
	cpm	%	cpm	%
Alcohol soluble proline	1689 (100)	56	340 (100)	94
Alcohol insoluble proline	9610 (100)	8	392 (100)	36
Total proline	11299 (100)	15	732 (100)	60
Concentrations of proline required for:				
1) Initial effect	10^{-6} M		10^{-6} M	
2) 1/2 Maximal effect	5×10^{-5} M		5×10^{-5} M	
3) Maximal effect	5×10^{-4} M		5×10^{-4} M	

Protocol as described in Table 5, (Fig. 3) (adapted from reference 38).

[a] Values obtained for proline ^{14}C in the absence of exogenous proline are in cpm per 40 root sections. ^{14}C values for proline in the presence of proline are presented as a percentage of the value obtained in the absence of exogenous proline.

proline biosynthesis which is insensitive to inhibition by proline and which becomes the dominant form of the enzyme as cells age. Bryan's group[42,50] has good evidence, at the enzyme level, that the regulation of homoserine dehydrogenase by effector molecules is altered as tissues age. The modification can be reproduced *in vitro*.[50] Thus the altered sensitivity to end product inhibiton seen *in vivo* could result from an altered proportion of isoenzymes,[47] from a modification of an already existing enzyme,[50] or from an altered distribution of enzyme and end product within the cell.[39]

In many tissue culture systems variant lines which are resistant to excess levels of a particular amino acid or to amino acid analogues have been identified.[48,51-54] They are usually dominant, are usually caused by a loss in the regulatory properties of a key enzyme, and usually result in an accumulation of particular end products. Recently there have been reports from several laboratories that these characteristics can be transferred to whole plants regenerated from cells in tissue culture.[48,53,54] These mutations, which have been induced in an *in vitro* situation, may represent the selection of cell types that already display or have the potential to display these characteristics as a result of a normal developmental sequence.

From all the work done to date it is apparent that amino acids can regulate amino acid biosynthesis, particularly in young cells. Thus amino acids or peptides supplied by the transport system can potentially reduce the flow of carbon from glucose or sucrose that would, in a nonregulated system, have gone into the synthesis of amino acids.

REGULATION OF THE ASSIMILATION OF NITRATE

Most plant cells, at most stages of development, have the capacity to reduce NO_3^- to the level of NH_4^+.[55] The two enzymes involved, nitrate and nitrite reductase, are induced by the primary substrate, NO_3^-. The induction may be modified by the availability of glucose or sucrose[13-15] and by the availability of amino acids, the potential end products of NO_3^- and NH_4^+ assimilation.[56-58] One of the problems at the moment is that it is difficult to determine the actual *in vivo* level of nitrate reductase. In seedlings or roots, for example, one can easily demonstrate nitrate reductase activity with either *in vitro*[13,59,60] or *in vivo*[13,61] assay methods and the prevention of the appearance of nitrate reductase activity by the addition of amino acids.[60,61] If one assays nitrate reductase by the reduction of $^{15}NO_3^-$ to $^{15}NH_4^+$ and ^{15}N-amino acids (that is, by a true *in vivo* assay) the activity is very low in seedling parts while the endosperm is supplying reduced nitrogen.[62,63] Under these conditions exogenous amino acids have only a minor effect on the reduction of $^{15}NO_3^-$.[64] Even when amino acids do have an effect on the induction of nitrate reductase,[60,61] the effective concentrations are in the mM rather than the μM range. Thus the actual *in vivo* importance of the amino acids in limiting the level of nitrate reduction in higher plants has not yet been established unequivocally.

In a converse test, when NO_3^- is added to the seedling system, there is no increase in the level of seedling protein relative to the minus NO_3^- treatment during the initial 4 days.[63,65,66] Thus there is apparently some constraint on the reduction or uptake of NO_3^- itself during the early post-germination stages. Although there is a clear reduction in the level of the free amino acids when the endosperm is removed,[10] this treatment alone is not sufficient to permit an enhanced induction of nitrate reductase in seedling leaves.[64]

Another method for determining the activity of nitrate reductase in seedlings or seedling roots is to examine the constituents of the "bleeding sap" in decapitated seedlings.[67,68] We have used this method to examine the capacity of the root system to reduce NO_3^--nitrogen and to transport that nitrogen to the shoot (Table 7). At 7 days the NO_3^- levels in the exudate are similar to NO_3^- levels in the medium, and NO_3^- accounts for at best about 50% of the total nitrogen in the bleeding sap. The total organic nitrogen is also significantly higher than the α-NH_2-nitrogen. This observation suggests that peptides or some unidentified nitrogen-containing compounds are a significant portion of the organic nitrogen content at this stage of development. By day 14 the α-NH_2-nitrogen and total organic nitrogen are clearly lower, suggesting that reduced nitrogen is no longer being supplied by the endosperm. The plants still lack the capacity to concentrate NO_3^- from the medium. However by 21 days the level of NO_3^- in the "bleeding sap" is much higher than the level in the medium. Thus once the endosperm reserves are exhausted there is a major shift in the capacity of the root to concentrate NO_3^- and to transport that NO_3^- to the shoot. The root at 21 days still has the capacity to reduce NO_3^- and that capacity is responsive to the external

Table 7. The effect of age on the components of the bleeding sap.

Age of seedlings	Concentration of NO_3^- in the nutrient medium	Concentration of bleeding sap constituents		
		NO_3^-	α-NH_2 N	"Organic N"
days	mM	mM	mM	mM
7	0	0.1	14.5	27.5
	1	1.0	12.8	25.0
	10	7.6	9.9	14.3
14	10	10.6	2.7	2.0
21	1	13.5	3.6	5.1
	10	22.6	5.5	5.9

Corn seedlings were grown in sand and vermiculite (1:20) for 7 day seedlings or in soil for older seedlings (14-21 days). They were watered with 0.1 Hoagland's which contained the indicated concentration KNO_3. NO_3^- was measured according to Cataldo et al,[69] α-NH_2 N according to Yemm and Cocking,[34] and total nitrogen by Kjehldahl digestion and Nesslerization.

concentrations of NO_3^-. The total organic nitrogen is only slightly higher than the level of α-NH_2-nitrogen which suggests that the transport of peptides or other unidentified nitrogen compounds is less important after the endosperm reserves have been exhausted. In the autonomous plant, however, most of the nitrogen moving to the top of the plant moves as NO_3^-, even when external sources of NO_3^- are low. Similar experiments have been performed with a maize mutant (opaque-12), which is deficient in nitrate reductase.[70] At 7 days when the endosperm is still an important source of nitrogen, the "bleeding sap" constituents are similar in mutant and wild-type siblings; at 14 days, however, the level of α-NH_2-nitrogen is lower in the mutant. One conclusion to be drawn from these experiments is that the supply of exogenous nitrogen is probably a relatively minor source of plant nitrogen while the endosperm is still actively supplying reduced nitrogen to the seedling. It appears that an active NO_3^--uptake system is absent at that stage.

SUMMARY

Proteins constitute 10 to 16% of the dry weight of the cereal endosperm. A prolamine fraction, which is located in protein bodies, and a glutelin fraction, which forms a structural matrix for the support of both protein bodies and amyloplasts, are the major proteins in the mature endosperm. They are substrates for the peptide hydrolases which develop in the endosperm in response to the endogenous levels of gibberellic and abscisic acids and to the addition of water. The products of hydrolysis (peptides and amino acids) are the major source of reduced nitrogen to the young seedling even when an external nitrogen source (NO_3^- or NH_4^+) is supplied.

The amino acids and peptides supplied by the endosperm suppress the biosynthesis of amino acids in the embryo. For example, in a seedling system, arginine, lysine, threonine, proline, valine, leucine, and isoleucine are regulated by end product inhibition. During early seedling growth, some factor appears to limit the availability of exogenous NO_3^- while the endosperm reserves are still available.

ACKNOWLEDGMENTS

I would like to thank B.M.R. Harvey, D.J. Mitchell, M.J. Winspear and R. Counts for the many experiments and discussions that contributed both results and ideas used in this paper.

REFERENCES

1. JACOBSEN JV, TJV HIGGINS, JA ZWAR 1979 Hormonal control of endosperm function during germination. In RL Phillips, CE Green, BG Gengenbach, eds, The Plant Seed: Development, Preservation and Germination, Academic, New York pp 241-262
2. TOOLE EH 1924 The transformation and course of development of germinating maize. Am J Bot 11: 325-350
3. OAKS A, H BEEVERS 1964 The glyoxylate cycle in maize scutellum. Plant Physiol 39: 431-434
4. FOLKES BF, EW YEMM 1958 The respiration of barley plants. X. Respiration and the metabolism of amino-acids and proteins in germinating grain. New Phytol 57: 106-131

5. JOY KW, BF FOLKES 1965 The uptake of amino acids and their incorporation into the proteins of excised barley embryos. J Expt Bot 16: 646-666
6. OAKS A 1963 The control of amino acid biosynthesis in maize roots tips. Biochim Biophys Acta 76: 638-641
7. OAKS A 1965 The effect of leucine on the biosynthesis of leucine in maize root tips. Plant Physiol 40: 149-155
8. SODEK L, CM WILSON 1970 Incorporation of leucine-^{14}C and lysine-^{14}C into protein in the developing endosperm of normal and opaque-2 corn. Arch Biochem Biophys 140: 29-38
9. SODEK L, CM WILSON 1973 Metabolism of lysine and leucine derived from storage protein during the germination of maize. Biochim Biophys Acta 204: 353-362
10. OAKS A, H BEEVERS 1964 The requirement for organic nitrogen in *Zea mays* embroys. Plant Physiol 39: 37-43
11. SHERIDAN WF, MG NEUFFER 1981 Maize mutants altered in embryo development. Symp Soc Dev Biol 39: 137-156
12. CAMERON-MILLS V, CM DUFFUS 1980 The influence of nutrition on embryo development and germination. Cereal Res Commun 8: 143-149
13. ASLAM M, A OAKS 1976 The effect of glucose on nitrate reductase in corn roots. Plant Physiol 56: 634-639
14. SAHULKA J, L LISA 1978 The influence of sugars on nitrate reductase induction by exogenous nitrate or nitrite in excised *Pisum sativum* roots. Biol Plant 20: 359-367
15. HÄNISCH TEN CATE CH, H BRETELER 1981 Role of sugars in nitrate utilization by roots of dwarf bean. Physiol Plant 52: 129-135
16. BARNARD RA, A OAKS 1970 Metabolism of proline in maize root tips. Can J Bot 48: 117-124
17. OAKS A, I STULEN, KE JONES, MJ WINSPEAR, S MISRA, I BOESEL 1980 Enzymes of nitrogen assimilation in maize roots. Planta 148: 477-484
18. WALL JS, JW PAULIS 1978 Corn and sorghum grain protein. Adv Cereal Sci Technol 2: 135-219
19. HARVEY BMR, A OAKS 1974 The hydrolysis of endosperm proteins in *Zea mays*. Plant Physiol 53: 453-457
20. OAKS A 1975 The regulation of nitrogen loss from maize endosperm. Can J Bot 43: 1077-1082
21. FELLER UT, T SOONG, RH HAGEMAN 1978 Patterns of proteolytic activities in different tissues of corn. Planta 140: 155-162

22. WINSPEAR MJ 1981 Peptide hydrolases in maize endosperm. MSc Thesis, McMaster University, pp 1-113
23. MIKOLA J 1982 Proteinases, peptidases, and inhibitors of endogenous proteinases in germinating seeds. In JG Vaughan, J Dussant, J Mosse, eds, Seed Proteins, Academic, London, New York, in press
24. HARVEY BMR, A OAKS 1974 Characteristics of an acid protease from maize endosperm. Plant Physiol 53: 449-452
25. ABE M, S ARAI, M FUJIMARI 1977 Purification and characterization of a protease occurring in the endosperm of germinating corn. Agric Biol Chem 41: 893-899
26. OAKS A, MJ WINSPEAR, S MISRA 1982 Hydrolysis of endosperm protein in Zea mays (W64A x W182E). In J Kruger, D LaBerge, eds, Third International Symposium on Pre-harvest Sprouting, Westview Press, Boulder, CO, in press
27. HARVEY BMR, A OAKS 1974 The role of gibberellic acid in the hydrolysis of endosperm reserves in Zea mays. Planta 121: 67-74
28. MAC GREGOR AW 1982 Cereal α-amylases: synthesis and action pattern. In JA Vaughan, J Dussant, J Mosse, eds, Seed Proteins, Academic, London, New York, in press
29. VARNER JF, DTH HO 1976 The role of hormones in the integration of seedling growth. In J. Papaconstantinow, ed, The Molecular Biology of Hormone Action, Academic, New York pp 173-194
30. SAWHNEY R, JM NAYLOR 1979 Dormancy studies in seed of Avena fatua. 9. A demonstration of genetic variability affecting the response to temperature during seed development. Can J Bot 57: 59-63
31. HARA I, K WADA, H MATSUBARA 1976 Pumpkin (Cucurbita sp) seed globulin. II. Alterations during germination. Plant Cell Physiol 17: 815-823
32. HARA I, H MATSUBARA 1980 Pumpkin (Cucurbita sp) seed globulin. V. Proteolytic activities involved in globulin degradation in ungerminated seeds. Plant Cell Physiol 21: 219-232
33. BAUMGARTEN B, MJ CHRISPEELS 1977 Purification and characterization of vicilin peptidohydrolase, the major endopeptidase in the cotyledons of mung-bean seedlings. Eur J Biochem 77: 223-233
34. YEMM EW, EC COCKING 1951 The determination of amino acids with ninhydrin. Analyst 80: 209-213

35. STEWARD CR, SF BOGGESS, D ASPINALL, LG PALEG 1977 Inhibition of proline oxidation by water stress. Plant Physiol 59: 930-932

36. HIGGINS CF, JW PAYNE 1977 Peptide transport by germinating barley embryos. Planta 134: 205-206

37. SAPONEN T 1979 Development of peptide transport activity in barley scutellum during germination. Plant Physiol 64: 570-574

38. OAKS A, DJ MITCHELL, RA BARNARD, FJ JOHNSON 1970 Regulation of proline biosynthesis in maize roots. Can J Bot 48: 2249-2258

39. OAKS A, RGS BIDWELL 1970 Compartmentation of intermediary metabolites. Annu Rev Plant Physiol 21: 43-66

40. BRYAN JK 1976 Amino acid biosynthesis and its regulation. In J Bonner and JE Varner, eds, Plant Biochemistry, Academic, New York pp 525-597

41. MIFLIN BJ, PJ LEA 1979 Amino acid metabolism. Annu Rev Plant Physiol 28: 299-329

42. BRYAN JK 1980 Synthesis of the aspartate family and branched chain amino acids. In BJ Miflin, ed, The Biochemistry of Plants, Vol 5, Amino acids and derivatives, Academic, New York pp 403-452

43. CARLSON PS 1970 Induction and isolation of auxotrophic mutants in somatic cell cultures of *Nicotiana tabacum*. Science 168: 487-489

44. KING J, RB HORSCH, AD SAVAGE 1980 Partial characterization of two stable auxotrophic cell strains of *Datura innoxia* Mill. Planta 149: 480-484

45. GILCHRIST GT, TS WOODIN, ML JOHNSON, T KOSUGE 1972 Regulation of aromatic amino acid biosynthesis in higher plants. I. Evidence for a regulatory form of chorismate mutase in etiolated mung bean seedlings. Plant Physiol 49: 52-57

46. CARLSON JE, JM WIDHOLM 1978 Separation of two forms of anthranilate synthetase from 5-methyltryptophan-susceptible and -resistant cultures of *Solanum tuberosum* cells. Physiol Plant 44: 251-255

47. SAKANO K 1979 Derepression and repression of lysine-sensitive aspartokinase during *in vitro* culture of carrot root tissue. Plant Physiol 63: 583-585

48. BRIGHT SWJ, BJ MIFLIN, SE ROGNES 1982 Threonine accumulation in the seeds of barley mutants. Biochem Genet 20: 229-243

49. OAKS A 1975 Changing patterns of metabolism as root cells mature. Biochem Physiol Pflanz 168: 371-374

50. DI CAMELLI CA, JK BRYAN 1980 Comparison of sensitive and desensitized forms of maize homoserine dehydrogenase. Plant Physiol 65: 176-183
51. HEIMER YM, P FILNER 1970 Regulation of nitrate assimilation pathway of cultured cells. II. Properties of a variant line. Biochim Biophys Acta 215: 152-165
52. WIDHOLM JM 1976 Selection and characterization of cultured carrot and tobacco cells resistant to lysine, methionine and proline analogues. Can J Bot 54: 1523-1529
53. HIBBERT KA, CE GREEN 1982 Inheritance and expression of lysine plus threonine resistance selected in maize tissue culture. Proc Natl Acad Sci USA 79: 559-563
54. BOURGIN JP 1978 Valine-resistant plants from _in vitro_ selected tobacco cells. Mol Gen Genet 161: 225-230
55. BEEVERS L, RH HAGEMAN 1980 Nitrate and nitrite reduction. _In_ BJ Miflin, ed, The Biochemistry of Plants, Vol 5, Amino Acids and Derivatives, Academic, New York, pp 115-168
56. FILNER P 1966 Regulation of nitrate reductase in cultured tobacco cells. Biochim Biophys Acta 118: 299-310
57. OAKS A 1974 The regulation of nitrate reductase in suspension cultures of soybean cells. Biochim Biophys Acta 372: 122-126
58. STEWART GR 1972 End product repression of nitrate reductase in _Lemna minor_ L. Symp Biol Hung 13: 127-135
59. OAKS A, W WALLACE, DL STEVENS 1972 Synthesis and turnover of nitrate reductase in corn roots. Plant Physiol 50: 649-654
60. OAKS A, M ASLAM, S BAKYTA 1977 Ammonium and amino acids as regulators of nitrate reductase in corn roots. Plant Physiol 59: 381-394
61. RADIN JW 1975 Differential regulation of nitrate reductase inductions in roots and shoots of cotton plants. Plant Physiol 55: 178-182
62. BLOHM D 1972 Untersuchungen zur Aminosäure-Biosynthese und Stickstoff-Assimilation in Keimpflanzenwurzeln. Diss Math Nat Fakultät, Humboldt-Universität, Berlin
63. SAMUKAWA K, M YAMAGUCHI 1979 Incorporation of ^{15}N-labeled inorganic nitrogen into amino acids in corn seedlings. Nippon Dojō Hyryō Gakkaishi (Japanese) 50: 323-326
64. OAKS A, I STULEN, I BOESEL 1979 The effect of amino acids and ammonium on the assimilation of $K^{15}NO_3$. Can J Bot 57: 1824-1829

65. SRIVASTAVA HS, A OAKS, IL BAKYTA 1976 The effect of nitrate on early seedling growth in Zea mays. Can J Bot 54: 923-929
66. DALE JE, GM FELIPPE, C MARRIOTT 1974 An analysis of the time of response of young barley seedlings to time of application of nitrogen. Ann Bot 38: 575-588
67. PATE JS 1973 Uptake, assimilation and transport of nitrogen compounds by plants. Soil Biol Biochem 5: 109-119
68. RUFTY TW, RJ VOLK, PR MC CLURE, DW ISRAEL, CD RAPER 1982 Relative content of NO_3^- and reduced N in xylem exudate as an indicator of root reduction of concurrently absorbed $^{15}NO_3^-$. Plant Physiol 69: 166-170
69. CATALDO DA, M HAROON, LE SCHRADER, RL YOUNGS 1975 Rapid colorimetric determination of nitrate in plant tissue by nitration of salicylic acid. Commun Soil Sci Plant Anal 6: 71-80
70. OAKS A, O NELSON 1982 Maize mutants deficient in nitrate reductase. Plant Physiol 69(suppl): 112

Chapter Five

AMINO ACID INTERCONVERSION IN GERMINATING SEEDS

P. J. LEA

Biochemistry Department
Rothamsted Experimental Station
Harpenden, Herts., England

K. W. JOY

Biology Department and Institute of Biochemistry
Carleton University
Ottawa, Ontario, Canada K1S 5B6

INTRODUCTION

The result of successful germination is the establishment of an autonomous seedling plant. Since the processes occurring in the seed are closely related to processes in the developing seedling, we have included a consideration of events in the young seedling in this discussion.

Plant seeds contain a high concentration of nutrient reserves which provide the raw materials for the growth of the seedling. The nitrogen reserve is mainly in the form of protein; quite high levels may be stored (30-40% dry weight) making such seeds of significant economic importance.

During germination there is a considerable interconversion of amino acids, necessitated by other metabolic events, the requirements of the transport system and differences between amino acid composition of reserves compared with new cytoplasmic protein. In contrast to the more balanced composition of average cytoplasmic protein in the seedling, the seed proteins often have a great predominance of a few amino acids.[1,2] In the protein hordein (barley), glx plus pro account for 60% of the amino acid residues,[1] and glx, leu, ala and pro make up a similar amount of the corn endosperm;[2] over half of the protein nitrogen in legumes such as pea or soybean is in the form of glx, asx, arg and leu;[3] see also Table 1. (Glx, asx, represents the total glutamate plus glutamine or aspartate plus asparagine residues, which are determined together as the free acid in protein hydrolysates). On the other hand, concentrations of some amino acids in seed proteins are lower than the amount required for protein synthesis,[2] and may also limit the dietary usefulness of these seeds as a total source of protein (sulphur-containing amino acids in legumes, lysine in corn).

Table 1. Comparison of amino acid compositions of cotyledon exudates, soluble pools of storage and seedling tissue, and storage protein. Expressed as molar percentage of total. Exudate and soluble samples taken after 4.5 days (mung bean) or 5 days of germination.

	Castor bean[5]		Pumpkin[10]				Mung bean[11]	
	Protein reserve	Exudate sol AA	Globulin protein	Cotyledon sol AA	Exudate sol AA	Hypocotyl sol AA	Cotyledon sol AA	Exudate sol AA
Asp	(11.4)	tr*	(10.0)	3.2	6.4	1.1	4.9	4.2
Asn		4.8		4.0	3.2	3.4	22.3	31.3
Glu	(19.2)	0.6	(20.7)	8.9	4.1	1.2	3.8	3.8
Gln		40.0		15.4	18.0	24.3	7.0	2.0
Thr	3.7	2.4	3.4	1.5	2.5	2.3	4.7	4.0
Ser	7.3	4.0	7.5	3.8	11.5	11.8	6.7	3.8
Gly	9.1	0.7	8.0	2.0	10.0	6.5	0.5	0.2
Ala	8.0	1.7	7.0	2.3	10.0	22.6	2.8	2.1
Cys	0.7	nd	0.7	tr	tr	tr	nd	nd
Met	1.4	tr	1.5	0.8	tr	0.6	0.4	nd
Ile	4.0	5.6	3.8	1.1	1.2	2.2	7.4	4.8
Val	6.0	11.3	5.8	2.7	2.7	3.3	6.3	8.9
Leu	5.6	3.3	8.4	8.1	1.8	3.0	7.0	6.7
Pro	5.4	4.0	0.4	3.2	1.9	1.0	4.1	5.9
Tyr	1.8	tr	1.5	3.9	tr	1.5	2.8	2.0
Phe	3.2	2.4	4.6	4.5	1.7	8.8	9.2	5.7
His	2.3	3.1	1.8	3.0	2.2	1.7	7.1	4.6
Lys	3.2	2.9	1.5	1.4	2.5	0.4	0.1	4.8
Arg	9.2	6.3	13.3	26.7	0.7	2.1	2.7	4.8
GAB*	nd*	2.5	nd	2.6	5.1	0.9	nd	nd
Orn	nd	nd	nd	0.9	14.4	1.2	nd	nd

* Abbreviations: GAB - γ-aminobutyrate; nd - not detected or not determined; tr - trace.

The metabolic changes in germination have been studied for many years. Some of the earliest rigorous plant biochemistry was carried out in the nineteenth century by Schulze, Prianischnikov and others who investigated the changes in nitrogenous components of germinating seeds. This classic work is extensively discussed by Chibnall.[4] Mobilization of the storage protein involves proteolysis (see Chapter 5 by A. Oaks); amino acids may then undergo considerable interconversion, leading in particular to synthesis of amides. Following transport of a selected group of amino acids and amides to the growing seedling, further interconversion is required to provide the full range of amino acids and nitrogenous components necessary for protein synthesis and growth. Some major processes are summarized in Fig. 1, and are discussed in detail below.

MAJOR PATHWAYS OF INTERCONVERSION

The Fate of Stored Amino Acids

Amino acids released from storage protein can contribute directly to protein synthesis in growing parts,[2] but labeling studies have shown that considerable interconversion and metabolism of amino acid carbon occurs in storage tissue during germination. Stewart and Beevers[5] supplied ^{14}C-amino acids to excised castor bean endosperm. For asp, glu and ala there was extensive metabolism to non-amino compounds (sugars, organic acids and CO_2 contained 30-70% of the original label after 2 h) and even within the amino acid fraction a relatively small proportion remained as the originally labeled substance. There was also considerable metabolism of labeled γ-aminobutyrate, gly and ser, and lesser metabolism of leu, gln, pro, val, arg and phe. In pea cotyledons,[6] glu and asp were extensively metabolized and there was substantial utilization of the carbon skeleton of leucine; a surprising observation was that even during a period of massive protein breakdown in the cotyledons, some amino acids were re-incorporated into the insoluble fraction. Other workers confirmed these results in peas, using leucine from which carbon was distributed mainly to organic acids, other amino acids and insoluble protein.[7] Leu and lys, derived directly from stored proteins, contributed to respiratory CO_2 in corn, as well as entering other metabolic compounds.[2] Work of this type and much analytical data shows that amino acids may contribute large amounts of substrate carbon to the respiratory system, and even to

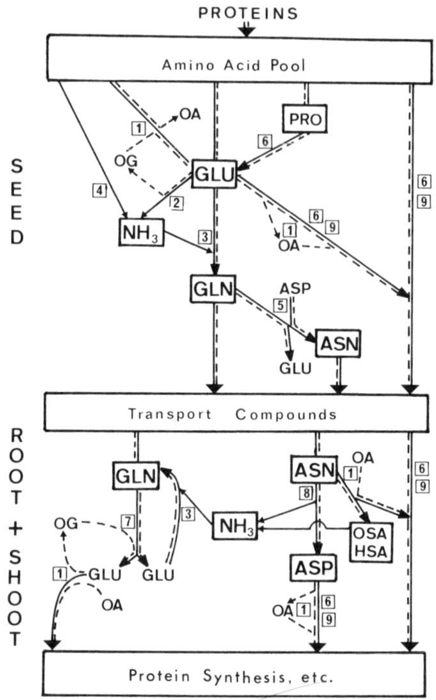

Fig. 1. Some central processes of nitrogen interconversion in germinating seedlings. OA - total pool of oxo-acids; OG - 2-oxoglutarate; OSA - 2-oxosuccinamate; HSA - 2-hydroxysuccinamate. Enzyme reactions and interconversions: 1 - aminotransferases; 2 - glutamate dehydrogenase; 3 - glutamine synthetase; 4 - specific deaminations (e.g. arg, urea, thr); 5 - asparagine synthetase; 6 - direct interconversions of amino acid skeletons; 7 - glutamate synthase; 8 - asparaginase; 9 - direct transfer without interconversion. Solid line - path of N; dashed line - path of C skeleton.

sugar synthesis in gluconeogenic seeds such as castor bean.[5] However it is possible that equilibration of label by transamination may over-emphasize the extent of net conversion in experiments of this type (see below).[8] It has been suggested that amino acids rapidly provide keto acids (which are not stored) to allow establishment of respiratory cycles in early stages of germination.[9] The diversion of amino acid carbon into respiration, etc., is accompanied by a release of ammonia which is reassimilated (and thus detoxified) then stored and transported as high nitrogen compounds, particularly the amides. There may also be demands for specific amino acids for synthesis of other nitrogenous materials, such as nucleic acids.

Regardless of these other demands, a substantial proportion of the reserve amino acids must be used in the synthesis of new seedling protein, although not necessarily in the form in which they are released. Analysis of soluble pools of storage tissue and growing regions provides some information about interconversions which are in progress and allow some assumptions to be made about the actual compounds transported from the seed. The work of Schulze and later workers[4] demonstrated the high accumulations of amides (asn, gln or both) to be found in germinating seedlings, and these high nitrogen compounds have been assumed to play a considerable role in transport. Some caution is in order, however, particularly when conclusions are drawn from tissues stressed by prolonged darkness. Smaller amounts of amino acids released from the reserves may be transported and used in protein synthesis yet may never accumulate to high levels.

Morphologically, growing seedlings do not provide an easy system for sampling the fluids which are transported to growing regions, presumably in the phloem. Exudates have been obtained from cotyledons of castor bean, pumpkin and mung bean. A comparison of the composition of exudate with the soluble amino acids of the cotyledon (Table 1) shows a reasonable similarity between accumulated and transported compounds, although some selectivity of transport clearly exists. The table also contrasts the composition of storage protein with that of the transported material, emphasizing the interconversions which occur. In the exudates most of the amino acids were present, many at low levels. In castor bean,[5] the major components were gln (40%), val (11%) and arg (6%); for pumpkin:[10] gln (18%), ornithine (14%), and ser, gly, ala (all about 10%); for mung bean[11] asn (31%) was the major transported compound.

Less commonly, other amino acids, often with high nitrogen content, may accumulate supplementing or replacing the amides, and may be involved in transport. This can be seen from Table 1, where ornithine has a role in pumpkin. Homoserine is the predominant amino acid in pea seedlings,[6] and in older plants remained a major transport compound in xylem and phloem.[12] Arginine in pine,[13] and γ-methylene glutamine in peanut[14] may also accumulate to high levels during germination.

Amino Acid Breakdown

Several types of reaction can be involved in the breakdown of amino acids; their catabolism has been reviewed by Mazelis.[15] Some may be metabolized so that they are converted directly into other amino acids of the same family by well established mechanisms (for example, pro to glu, also see below). However, for many amino acids the process of catabolism involves removal of nitrogen from the carbon skeleton, which then undergoes breakdown or interconversion. The fate of the carbon skeleton may be to provide the basis for an alternate amino acid, a respiratory substrate, or other non-nitrogenous metabolic components such as keto acids or even sugars. In all but the first case there will be a surplus of nitrogen.

Two main types of reaction are involved in removal of nitrogen from simple amino acids, namely transamination and deamination. Transamination results in transfer of the amino group to an alternate keto acid, and gives no net change in total amino acid, whereas deamination removes the nitrogen from organic form producing ammonia which must then be reassimilated in amide synthesis. More elaborate and specific pathways are involved for complex amino acids, for example arginine is first cleaved to give ornithine and urea, ornithine is further metabolized to produce glu, and urea is broken down into ammonia plus carbon dioxide.[15]

Transamination. Transaminases (aminotransferases) carry out the following exchange reactions between α-amino acids and keto acids:

$$R_1CHNH_2COOH + R_2COCOOH \rightleftharpoons R_1COCOOH + R_2CHNH_2COOH$$

With a few reported exceptions, transaminases are freely reversible, and this allows equilibration of carbon between an amino acid and its corresponding keto acid, even though no net flow occurs.[8] This may lead to misleading results in ^{14}C-labeling experiments, over-emphasizing the extent of metabolism.

There have been many studies on transaminases in seeds and seedlings, and some are discussed in a review by Wightman and Forest.[16] Wilson et al[17] showed that preparations from lupin and barley seedlings would catalyze the transamination of 17 amino acids with α-ketoglutarate. Forest and Wightman[18] measured the activity of 22 different amino acids for transamination to α-ketoglutarate, oxaloacetate, pyruvate and glyoxylate in extracts from seedling and cotyledon tissues of Phaseolus. Proline, hydroxyproline and cystine were not transaminated, while serine reacted only with pyruvate and glyoxylate. All other protein amino acids underwent transamination with the four keto acids. Ala-glu and asp-glu activities were greatest. In a related study, the same authors showed that the major transaminase activities were initially high in germinated cotyledons, but decreased with time, while activities generally increased with time in seedling shoot and root tissue; data for changes in soluble amino acids were also reported.[19] Transamination appears to be the major route for catabolism of asp, ala, gly, ile, leu, ser, val and the aromatic amino acids.[15] Much of the nitrogen harvested in this way will flow into glu; from there it may be redistributed by further transamination, or released as ammonia by deamination.

Deamination. Several enzymes are capable of removing an amino group to release free ammonia, including L-amino acid oxidases, deaminases and dehydrases,[20] but as far as we are aware, these enzymes have not been detected in plants, or at least not in germinating seeds.

Glutamate dehydrogenase is of very widespread distribution in plant tissues, and carries out the reaction

$$\text{Glu} + \text{NAD(P)}^+ + H_2O \rightleftharpoons \text{2-oxoglutarate} + \text{NAD(P)H} + NH_4^+$$

Since the breakdown of many amino acids results in production of glu directly, or through transamination, it appears likely that this is the main route for ammonia release during seed germination. Glutamate dehydrogenase has been detected in seeds or seedlings of many plants, although in a number of cases there has been no attempt to differentiate between the enzyme of seed reserve as opposed to growing seedling tissue. A partially purified enzyme from soybean cotyledons has been well characterized.[21] It has been shown that isoenzymes of glutamate dehydrogenase exist in pea and bean (Vicia) seeds and seedlings,[22] pumpkin cotyledons,[23] and seedlings of barley, corn, rice, mung bean, groundnut and *Vigna*.[24] A comprehensive review of the properties of glutamate dehydrogenase is given by Stewart et al.[25]

A consideration of the amino acid changes occurring and the properties of the enzyme (particularly the high K_m for ammonia[21]) suggest that in seed tissue glutamate dehydrogenase must have a deaminating role; a role in glutamate synthesis would result in useless cycling of glutamate and would be of no benefit. It is possible that in seedling tissue the enzyme could have an assimilatory role, although this may be unlikely if glutamine synthetase, with a high affinity for ammonia, is present--see below.

Amide Synthesis

Glutamine synthetase. Much information exists about this enzyme in plants,[25] and it is thought to play a predominant role in ammonia assimilation.[26,27] It has been detected in many tissues, including pea seeds,[28] and mung bean cotyledons.[11] It catalyzes the reaction:

$$\text{L-glu} + NH_3 + \text{ATP} \longrightarrow \text{L-gln} + \text{ADP} + P_i$$

The enzyme catalyzes several other synthetic or transfer reactions, and may also act as a glutaminase in the absence of ATP.

It is likely that the major proportion of the ammonia, released in the interconversion and deamination of seed amino acids, is reassimilated through glutamine synthetase. This seems to be true also for seeds which accumulate and export large amounts of asparagine, since glutamine is the favored amide donor in asparagine synthesis; in mung beans, in which asparagine is the main export compound leaving the cotyledons, glutamine synthetase is present in the dry seed and increases during the first few days of germination.[11]

Asparagine synthetase. Asparagine is the most abundant amino acid in many germinating seeds, necessitating a considerable synthesis. Asn synthetase has been found in many seed reserve tissues, which are in fact the major location where this enzyme has been detected.

Although original work suggested that in lupine, ammonia provided the amide group,[29]

$$\text{Asp} + \text{ammonia} \xrightarrow{\text{ATP}} \text{Asn}$$

later work indicated that glutamine was a preferred amide donor in lupin,[30,31] mung bean[11] and soybean,[32]

$$\text{Asp} + \text{gln} + \text{ATP} \longrightarrow \text{Asn} + \text{glu} + \text{AMP} + \text{PP}_i$$

The affinity for gln is 20-fold (or more) greater than for ammonia and assays show a 3-fold or greater maximum activity with gln (see review by Lea and Miflin[33]). However this does not rule out the possibility that some ammonia may flow directly into asn, rather than flowing by way of gln.

In mung beans the enzyme is synthesized in the cotyledons during germination; the activity may be regulated by an inhibitor produced by the seedling.[11]

Amide Utilization

Amides arriving in the developing seedling will be required in small amounts for direct incorporation into new protein, and gln will be used in nucleotide biosynthesis, but the amides also provide an abundant high nitrogen source for synthesis of amino acids. The specific problem involved in this conversion is the need to transfer an amide group into the α-amino position.

Utilization of glutamine. Many tissue extracts exhibit a general glutaminase activity, which results in release of the amide group of gln as ammonia, but this activity is not well characterized and can be shown by enzyme systems such as gln synthetase[23] and asn synthetase.[30] This would not be an appropriate system for utilization of gln-amide, since reassimilation of the released ammonia should occur through gln synthetase, and would lead to futile cycling.

The major pathway for utilization of gln appears to be through glutamate synthase (GOGAT) in which the amide group is directly utilized to give an increase in amino nitrogen:

$$\underline{\underline{\text{L}}}\text{-gln} + \text{2-oxoglutarate} + 2[\text{H-donor}]_{\text{reduced}} \longrightarrow 2\ \underline{\underline{\text{L}}}\text{-glu} + 2[\text{H-donor}]_{\text{oxidized}}$$

In photosynthetic tissue, ferredoxin is the source of reductant, but NAD(P)H may also be involved, particularly in non-green plant parts; see review by Stewart et al.[25] The enzyme has been found in a number of plant tissues,[25] but as far as we are aware there are no reports of its presence in the germinating seed (although it is present in developing seeds); this is not a cause for concern, since there is no requirement for conversion of gln to glu in the storage region. The enzyme, both Fd and NADH dependent forms, is present in the roots and shoots of developing pea seedlings.[34]

Utilization of asparagine. Metabolism of asn has been reviewed recently by Lea and Miflin.[33] So far, there have not been any definitive reports of an amidotransferase system, analogous to glu synthase, which can utilize asn. It has been suggested that such activities reported for glu synthase preparations were the result of slight reagent impurities.[33] At present, it appears that utilization of the amide group of asn for amino acid synthesis requires release as ammonia, and reassimilation via gln synthetase and glu synthase. Considering that many plants transport so much of their nitrogen as asn, this circuitous route appears to be rather inefficient.

Asparaginase provides one route for the removal of amide nitrogen:

$$Asn + H_2O \longrightarrow Asp + NH_3$$

In earlier work, the enzyme was found in a restricted range of plants, or specialized parts such as root nodules,[35] but it was later shown to be more widespread when it was realized that in many species the enzyme is strongly K^+-dependent.[36] It has been shown to be present in developing pea leaves, but rapidly declines while the leaf approaches full expansion.[37]

Labeling experiments[38] suggested that asn nitrogen (^{15}N) was also rapidly used in amino acid synthesis by transamination, since the transfer of label occurred much more extensively than suggested by the labeling of free ammonia or asp (which would result from asparaginase activity). This has been confirmed by use of asn ^{15}N-labeled specifically in the amino position, which is rapidly found in the ala, glu and homoserine of growing pea leaves (K. W. Joy, T. C. Ta and R. J. Ireland, unpublished). Asn transaminase activity was noted in lupin and barley,[17] and later in soybean leaves[35] and peas.[37] Lloyd and Joy[39] showed that in growing pea leaves there was utilization of asn by tranamination and the product which accumulated was not the direct transamination product, 2-oxosuccinamic acid, but a reduced derivative, 2-hydroxysuccinamic acid. Other evidence suggests that the transaminase pathway is the predominant system operating in pea leaves, although asparaginase is more important in the developing seeds.[37] Operation of the transaminase route still does not solve the problem of utilization of the amide group of asn. Deamidation of oxosuccinamate has been shown in soybean leaf extracts,[35] and deamidation may also account for the metabolism of hydroxysuccinamate.[39]

METABOLISM OF SOME SPECIAL AMINO ACIDS

Glutamate Family--Arginine and Proline Metabolism

These two amino acids will be considered together as they are both initially derived from glutamate and there is some evidence that arginine may be converted directly to proline under conditions of stress.

Arginine: Certain seeds (e.g., *Cucurbita*,[10] *Vicia*[40,41] and *Dolichos*[40]) contain storage proteins with high levels of arginine. Upon germination, the amino acid is rapidly metabolized to ornithine, urea and glutamate via reactions 1-4 (Fig. 2). Arginase increases dramatically during the early stages of development of seedlings of all three plants mentioned above,[40,42] along with urease which is required to

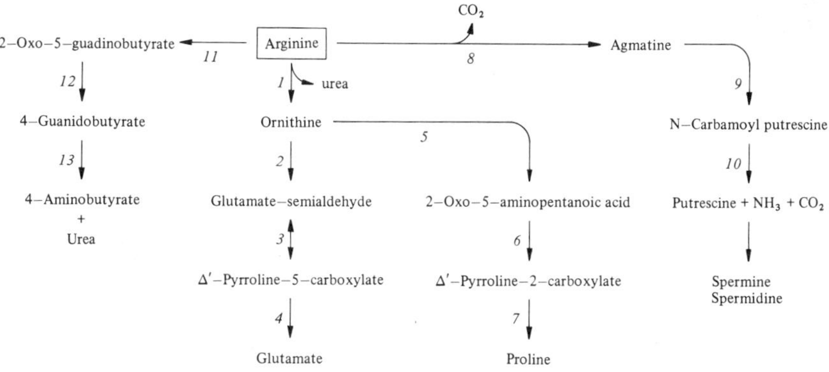

Fig. 2. Possible routes of arginine metabolism in germinating seeds (reproduced with permission).[26]

convert urea to ammonia and CO_2.[43,44] The ammonia is then reassimilated to glutamine, which along with ornithine makes up the two most important nitrogen transport compounds of the pumpkin seed (Table 1).[10]

Canavanine (an oxygen containing analogue of arginine) is present in very high concentrations in seeds of *Canavalia ensiformis*. On germination the compound is metabolized in a very similar manner to arginine by the enzyme arginase to yield canaline and urea.[45,46] Arginine, which is a major transport compound in both deciduous[47] and coniferous trees,[48] may also be metabolized to polyamines[49] (route 8-10) and various guanido compounds[48] (route 11-13) as shown in Fig. 2.

In the majority of plants studied arginine is catabolized in the cotyledons during germination and not transported in any great quantity to the growing points. However, in pea seeds there is evidence that arginine is synthesized *de novo* by the classical route (1-10 in Fig. 3).[50] Carbamoyl phosphate synthetase (7),[51] ornithine carbamoyltransferase (8),[52,53] arginosuccinate synthetase (9)[54] and arginosuccinate lyase (10)[55] have all been studied in detail in pea seedlings.

Although seedlings may have both the capacity to synthesize and degrade arginine via ornithine (reactions 8-10, Fig. 3 and reaction 1, Fig. 2), it must be stressed that these reactions do not constitute the Krebs-Henseleit ornithine cycle that operates in animals as a mechanism of urea excretion. Such a cycle would be futile and energy wasting in plants as the ammonia evolved by urease would have to be reassimilated. It would appear that arginine is synthesized in one organelle (preliminary evidence suggests the chloroplast)[56] by a pathway that is tightly regulated, and is broken down either in a different tissue or at a different stage of development of the plant in the mitochondria. A similar case for the non-functioning of the Krebs-Henseleit cycle in canavanine metabolism can also be argued.[57]

Proline. Proline is a major constituent of specific storage proteins in cereals.[58] The imino acid will accumulate in seedlings during germination[59] particularly if grown in the dark (e.g., malting barley).[60] Normally the proline is converted to glutamate by the reversal of reactions 13, 12 and 11 in Fig. 3. However the conversion of proline to pyrroline-5-carboxylate is not carried out by the reversal of the enzyme

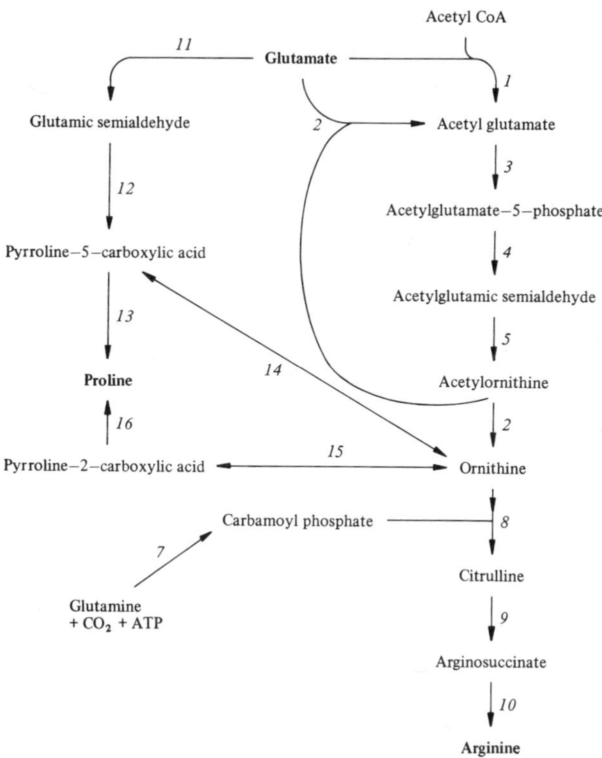

Fig. 3. The pathway of biosynthesis of arginine, and three possible routes of proline synthesis (reproduced with permission).[26]

pyrroline-5-carboxylate reductase, but by direct oxidation in the mitochondria.[61]

Proline synthesis also takes place in a number of germinating seedlings (e.g., Cucurbita[10]) and has been extensively studied in maize roots where feedback regulation alters as the root ages.[62] Evidence for the precise route of proline synthesis in germinating seedlings is limited. Most studies have been carried out during conditions of water stress when proline accumulates to very high levels.[63] Route 11-13 in Fig. 3 has been demonstrated in beet[64] and stressed barley.[65] Pyrroline-5-carboxylate reductase has been thoroughly examined[66,67] but the other two enzymes have not been studied. Labeling data suggests that arginine may be converted to proline via route 1-7 (Fig. 2).[68] Such a route would involve the transfer of the 2-amino group of ornithine and the formation of pyrroline-2-carboxylate. Evidence as to the ability of ornithine to transaminate either its 2-amino or 5-amino group is somewhat confused. Ornithine: 2-oxo acid 5-amino transferase (reaction 14, Fig. 3) has been claimed to have been measured in a number of plants,[69-71] but the assay technique was not specific for pyrroline-5-carboxylate; pyrroline-2-carboxylate also being a possible product. Evidence for the presence of an ornithine: 2-oxo acid 2-aminotransferase has been obtained[72,73] and pyrroline-2-carboxylate reductase has also been detected.[74,75]

Thus the precise route of proline synthesis in any plant tissue under normal as opposed to water stressed conditions has not been determined. Enzymological evidence suggests that any of these routes could operate. Unfortunately the existence of a maize mutant auxotrophic for proline[76,77] has not given any definitive evidence as to which pathway is functioning. Studies are presently being carried out at Rothamsted on a barley mutant that overproduces proline during the germination stages.[78]

Biosynthesis of the Aspartate Family of Amino Acids

The biosynthesis of the aspartate family of amino acids is shown in Fig. 4. The pathway was initially determined in bacteria, and has been shown to hold true in all plants so far studied by precursor-feeding and enzymological studies.[26] According to localization data obtained predominantly at Rothamsted, in green leaves, the synthesis of lysine, threonine, isoleucine and homocysteine takes place solely in the chloroplasts.[79-82] Presumably in non-green tissues the processes take place in plastids. The enzymes involved in the conversion of homocysteine to methionine and _S_-adenosylmethionine are apparently localized in the cytoplasm.[82]

Apart from the unusual case of _Pisum_, with synthesis of large amounts of homoserine (which will be discussed in a later section), there is little evidence that the amino acids in this pathway accumulate to any extent during germination or any other time during the plant growth. The pathway is feedback regulated by the three main products lysine, threonine and _S_-adenosylmethionine (Fig. 5). Lysine is a strong inhibitor of dihydrodipicolinate synthase (3) (EC 4.2.1.52)[83] and one of the isoenzymes of aspartate kinase

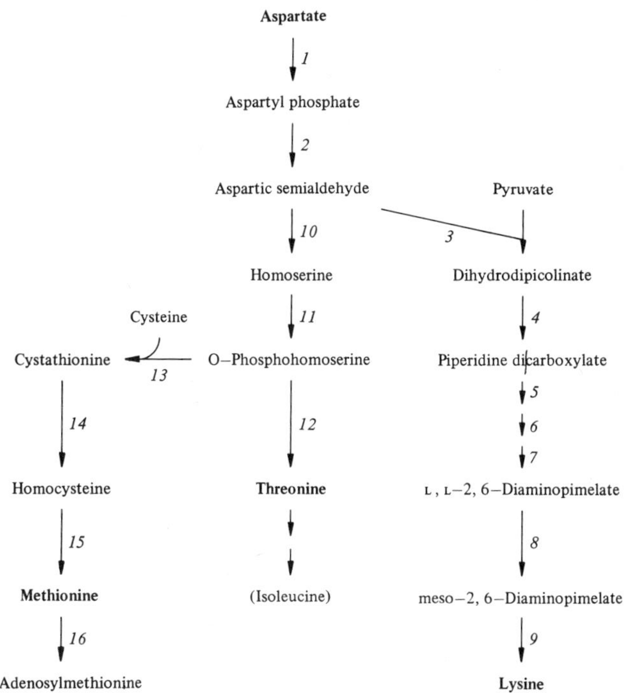

Fig. 4. The synthesis of amino acids derived from aspartate (reproduced with permission).[26]

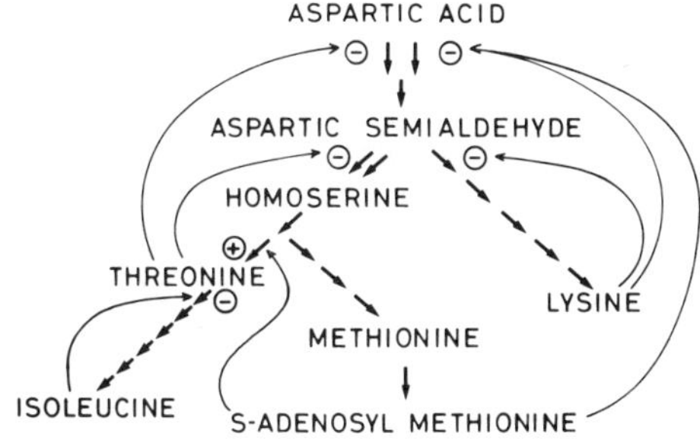

Fig. 5. Regulation of the aspartate pathway in higher plants. ⊖→ Inhibition ⊕→ Stimulation

(1) (EC 2.7.2.4).[84-87] S-adenosylmethionine is a synergistic inhibitor of aspartate kinase in the presence of lysine[88] and also activates threonine synthase (12) (EC 4.2.99.2)[89,90] in order to direct the flow of carbon away from methionine towards threonine. Threonine is an inhibitor of one isoenzyme of aspartate kinase[87,91] and one of homoserine dehydrogenase (10) (EC 1.1.1.3).[81,92-94] It could be predicted from the branched pathway that methionine or S-adenosylmethionine would also inhibit cystathionine synthase, (13) the first enzyme unique to methionine synthesis. Although this has not been demonstrated, there has recently been evidence to suggest that the level of methionine (or a product) may regulate the absolute amount of cystathionine synthase activity present in *Lemna*.[95]

Cereals (with special reference to Hordeum). A major method employed at Rothamsted for examining amino acid metabolism in barley has been the germination of isolated embryos on sterile agar plates, a technique known since 1890. It is immediately clear that barley embryos have the capability to synthesize from nitrate all the protein amino acids required to produce normal leaves and roots during germination. There is no absolute requirement for any amino acids from the endosperm provided an inorganic form of nitrogen is available.[96a] However, if amino acids are supplied to the germinating embryos, they are rapidly taken up and immediately switch off the appropriate section of the pathway.[96a,b]

Lysine and threonine when added together are known to inhibit the growth of a wide range of plants by preventing the synthesis of methionine.[97] Three mutant lines of barley have been selected that are resistant to the combined action of these end product amino acids and have been designated R2501, R3004 and R3202.[98-100] Separation of aspartate kinase (AK) by DEAE cellulose chromatography (Fig. 6) indicates there are three isoenzymes: I sensitive to threonine, and II and III sensitive to lysine and S-adenosylmethionine. The characteristics of AKI in all the mutants remained the same. However, in R2501, AKII was 25-fold less sensitive to lysine inhibition and in R3202, AKII was present at very low levels and was totally insensitive to lysine. In R3004, AKII was normal but AKIII was 10-fold less sensitive to lysine. Two of the mutant lines were shown to have significant increases in soluble threonine in the seed (Table 2), although none of the plants were shown to accumulate lysine, probably due to the strict control of dihydrodipicolinate synthase.

Crossing studies suggest that AKII and AKIII are under separate gene control; lt1 and lt2.[99]

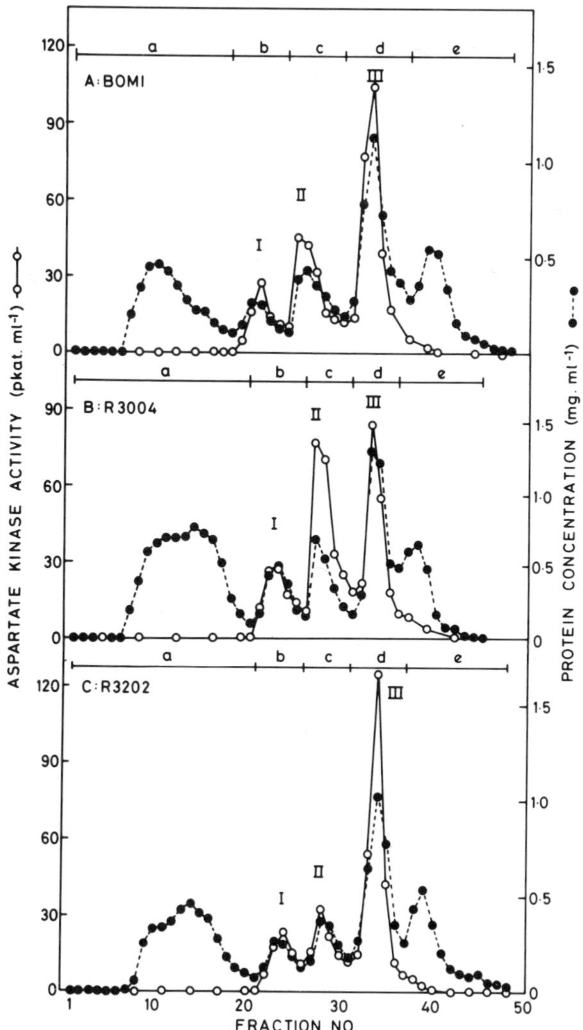

Fig. 6. Separation on DEAE-cellulose of three forms of aspartate kinase from wild-type barley (BOMI) and two mutant lines selected for resistance to lysine and threonine. ●---● - protein content; ○——○ - aspartate kinase activity (reproduced with permission).[100]

Table 2. Soluble amino acids in grains of normal barley and three mutants.

	Soluble content (nmol/mgN)				
Amino acid	Pooled controls I	II	R3004	R3202	R2501
Thr	9 ± 3	12 ± 5	116 ± 36	16 ± 5	147 ± 21
Lys	12 ± 5	12 ± 5	15 ± 3	18 ± 5	19 ± 8
Ala	49 ± 13	47 ± 10	41 ± 6	68 ± 14	29 ± 5
Val	18 ± 10	28 ± 12	17 ± 5	38 ± 10	19 ± 4
Glx	143 ± 25	140 ± 36	122 ± 20	199 ± 39	112 ± 22
Asx	167 ± 27	181 ± 28	122 ± 18	237 ± 45	131 ± 24
Total	580	740	745	905	743

Reproduced with permission from Bright et al.[99]

Neither the relative proportions of the isoenzymes of aspartate kinase, nor their sensitivity to feedback inhibitors alters in the leaves during the germination process, although the presence of increasing levels of ATPase activity tends to inhibit aspartate kinase activity in the older seedlings if it is not removed by DEAE-cellulose chromatography (P. Arruda, unpublished results). If the barley seedlings are grown on various concentrations of lysine, threonine or methionine there is no indication of any "repression" of the levels of aspartate kinase activity (S. E. Rognes, unpublished results), although this has been suggested to occur in tissue culture cells.[101] In germinating wheat[86] and maize,[102] aspartate kinase has also been shown to be sensitive to lysine but no further work has been carried out on the enzyme from these sources.

Homoserine dehydrogenase in barley seedlings exists in two readily separable forms.[81,85] A high molecular weight isoenzyme in the chloroplast is sensitive to threonine and a low molecular weight isoenzyme in the cytosol is sensitive to cysteine.[81] In maize, homoserine dehydrogenase has been extensively studied by Bryan and his colleagues[93,103-106] and has been shown to exist as a low molecular weight threonine resistant form (Class I) and a high molecular weight threonine sensitive isoenzyme designated class II/III due to the presence of interconvertible forms. In dark grown germinating seedlings the ratio of Class I:Class II/III isoenzyme is almost 1:6 and remains constant over 13 days of germination. In light grown seedlings however there is an alteration in the levels of the two forms of homoserine dehydrogenase as the leaf matures. At the basal meristematic regions there are high levels of Class II/III compared to Class I but this level falls during maturation. At the leaf tips the two isoenzymes were present at equal levels.[106] Unfortunately no information is available as to whether there is any alteration in the pattern of synthesis of threonine and methionine during the germination process.

In germinating maize and barley it was originally reported that the sensitivity of the Class II/III isoenzyme of homoserine dehydrogenase decreased with age.[81,103,104] However more recent evidence suggests that particularly in maize these results may have been an artefact of the extraction procedure.[92] The addition of high levels of the thiol reagent dithioerythritol to the extraction medium allows the isolation of large amounts of the threonine sensitive isoenzyme.[92] The work is a very good example of the precautions that must be taken to optimize the extraction and assay conditions of any enzyme during a developmental series of seedling germination. The occurrence of phenols, proteases and ATPases in increasing amounts in older tissues is well documented.

The synthesis of methionine and _S_-adenosylmethionine is not further feedback regulated, although recent evidence suggests that the level of cystathionine synthase in _Lemna_ may be "repressed" by the addition of methionine.[95] Preliminary data utilizing sterile barley embryos germinated on methionine confirms this finding (S. E. Rognes, unpublished results). It is interesting that the enzyme shows a sharp peak of maximal activity four days after the onset of germination in intact barley seedling leaves.[107] Homoserine kinase (11) (EC 2.7.1.39) and threonine synthase have been studied in germinating barley.[90] The latter enzyme is strongly stimulated by _S_-adenosylmethionine as was originally shown in _Beta vulgaris_.[89]

Dihydrodipicolinate synthase isolated from germinating wheat and maize is very sensitive to lysine inhibition, a 50% maximal effect occuring at 11 μM lysine.[83,108]

Legumes (with special reference to *Pisum*). Although homoserine is an intermediate in the biosynthesis of threonine, isoleucine and methionine in all organisms so far examined,[26] the amino acid only accumulates in the two closely related plant genera of *Pisum* and *Lathyrus*.[109] The accumulation during germination was initially demonstrated by Virtanen et al,[110] and confirmed by Lawrence and Grant[111] and Larson and Beevers.[112] A more recent demonstration of the massive *de novo* synthesis of homoserine following the onset of germination can be seen in the data of Van Egeraat[113] reproduced in Table 3. In this section we will try to explain this unusual alteration to the flow of carbon down the aspartate pathway, by comparing the enzymes of *Pisum* with those of other higher plants.

A number of labeling studies have indicated that aspartate is a precursor of homoserine in germinating peas.[112,114]

Table 3. Free ninhydrin-positive compounds (μg/plant) in cotyledons of uninoculated pea plants of different age, grown without added nitrogen.

Compounds	Age (days after wetting the seed)										
	0	2	3	4	5	6	7	10	18	21	24
'Unknown X'	—	—	—	—	—	—	—	—	—	—	—
'Unknown Y'	tr.	+	+	+	+	+	+	+	+	+	+
Aspartic acid	50	106	225	95	106	145	133	225	560	40	tr.
Threonine	tr.	48	42	42	36	36	36	119	320	105	tr.
Serine	tr.	168	220	273	220	325	357	378	662	115	tr.
Asparagine/ glutamine	90	830	620	680	680	740	830	880	1925	300	tr.
Homoserine	tr.	60	350	715	1150	3000	2511	550	298	48	tr.
Proline	tr.	tr.	tr.	35	35	70	35	35	23	tr.	tr.
Glutamic acid	265	840	660	700	735	760	748	1470	690	75	tr.
Glycine	23	34	30	30	30	75	75	100	200	45	tr.
Alanine	22	63	45	40	63	175	186	195	285	45	tr.
Valine	tr.	tr.	tr.	35	35	70	46	47	35	tr.	tr.
Cystine	tr.	84	144	310	335	360	408	600	1100	240	tr.
Methionine	tr.	tr.	tr.	tr.	tr.	45	30	30	30	tr.	—
Isoleucine	tr.	tr.	tr.	78	78	78	131	140	300	26	tr.
Leucine	tr.	tr.	tr.	52	52	78	118	140	390	tr.	tr.
Tyrosine	tr.	tr.	tr.	54	36	36	72	72	180	38	tr.
Phenylalanine	tr.	tr.	tr.	65	82	115	165	230	330	65	tr.
γ-Aminobutyric acid	tr.	tr.	tr.	tr.	tr.	tr.	tr.	56	61	30	tr.
Lysine	tr.	55	64	73	91	100	183	240	458	55	tr.
Histidine	tr.	46	40	62	93	186	155	230	420	110	tr.
Ammonia	34	24	41	58	90	53	68	60	44	44	tr.
Arginine	230	735	680	630	730	840	820	840	1050	210	tr.

However, a more detailed study by Mitchell and Bidwell suggested that aspartate was initially extensively metabolized probably via the tricarboxylic acid cycle prior to its conversion to homoserine.[115] Utilizing isolated intact chloroplasts from young pea seedlings Mills et al[79] were able to demonstrate rapid synthesis of [^{14}C]-homoserine from [^{14}C]-aspartate in the light, which was tightly regulated by lysine and threonine. Homoserine was also found to be present in pea chloroplasts but at a lower concentration than that of the cytoplasm.[116]

Two other routes of synthesis of homoserine in peas have been suggested: either from methionine[117] or by a direct transamination from asparagine.[38] These two possibilities remain to be confirmed.

Aspartate kinase from germinating pea cotyledons is predominantly threonine sensitive[91] and is similar to that found in soybean cotyledons. In the young leaf although a threonine sensitive enzyme is present, the enzyme is far more sensitive to lysine,[87] and is subject to cooperative inhibition by S-adenosylmethionine.[88]

In two investigations of pea leaf homoserine dehydrogenase a threonine sensitive form was clearly demonstrated.[81,91] No evidence was obtained of any alteration of the sensitivity during the germination of the seedlings. However, Di Marco and Grego[94] have suggested that the enzyme is only sensitive to threonine in the reverse direction and thereby allows homoserine to accumulate in the forward direction. Further studies confirmed the existence of this undirectional inhibition at different ages of pea leaves during germination.[119] The reasons for the discrepancies between the results of the three groups of workers is as yet not clear. In soybean a progressive decrease in the amount of inhibition of homoserine dehydrogenase by threonine occurs in the roots, stems and cotyledons during germination.[93] However, as neither dithiothreitol nor dithioerythritol was employed in the extraction medium, it is possible that the apparent desensitization is an artefact similar to that detected in maize.

Perhaps the most clear cut evidence for the reason for the accumulation of homoserine in peas, has been obtained by Aarnes and his colleagues. In pea, homoserine kinase has a K_m value for homoserine of 6.7 mM;[120] 17-fold higher than for the barley seedling of similar purity and 80-fold higher than the value of a highly purified barley leaf enzyme. Coupled with this in pea, isoleucine, valine and ornithine are potent competitive inhibitors of homoserine kinase, with a stronger affinity for the enzyme than homoserine itself. The K_i values of isoleucine and valine are low enough to exert a regulatory effect *in vivo*. In barley no such inhibition by amino acids occurs.[90]

Thus it is possible to envisage in pea seedlings a possible mechanism by which homoserine may accumulate. Homoserine synthesis initially takes place via a threonine sensitive aspartate kinase and homoserine dehydrogenase. Due to a high K_m for homoserine and inhibition by other amino acids, homoserine kinase functions only very slowly. Thus threonine synthesis should be depressed and it would not build up to a sufficient level to inhibit aspartate kinase and homoserine dehydrogenase; it has in fact been shown that there is a low level of threonine in pea chloroplasts.[116] If as has been suggested by enzyme localization studies these reactions are taking place in the chloroplasts (and plastids?) then homoserine may diffuse out of the organelle and accumulate in the cytoplasm.

Homoserine is not only a precursor of the protein amino acid threonine, methionine and isoleucine.[26,121] Various feeding studies to germinating seedlings have indicated that it is metabolized to α,γ-diaminobutyrate,[122] *O*-acetylhomoserine,[123] *O*-oxalylhomoserine,[124] azetidine-2-carboxylate,[125] canavanine[57] and 2-amino-4-(isoxalin-5-one-2-yl) butyric acid[126] (S. E. Rognes, unpublished results). The existence of a cytoplasmic homoserine dehydrogenase[81] indicates that the conversion of homoserine to aspartate semialdehyde may be the first step in this metabolism; a suggestion confirmed by the very recent finding that lysine became rapidly labeled after the feeding of ^{14}C-homoserine in *Canavalia*.[57] A tentative scheme of the possible routes of homoserine metabolism is shown in Fig. 7. It would be interesting to compare the kinetics of homoserine kinase from plants that are producers and non-producers of unusual amino acids, to ascertain whether a build-up of homoserine leads to a diversion of this amino acid into more readily stored end products.

CONCLUSIONS

Germination is a period of intense metabolic activity including many processes involved in breakdown, transport

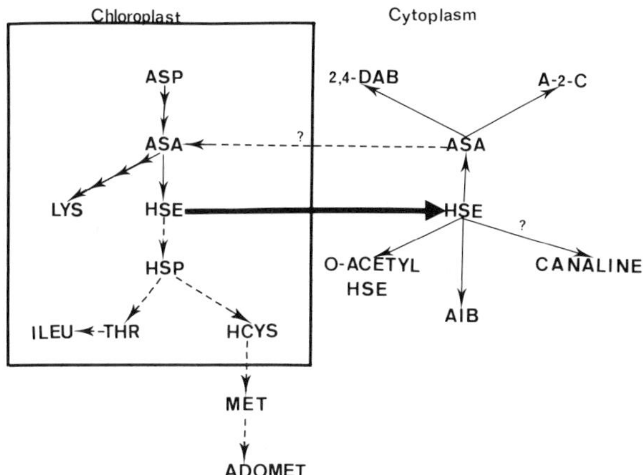

Fig. 7. Suggested pathways for metabolism of homoserine in various plant species. Abbreviations: ASP, aspartate; ASA, aspartate semialdehyde; HSE, homoserine; HSP, O-phosphohomoserine; LYS, lysine; THR, threonine; ILEU, isoleucine; HCYS, homocysteine; MET, methionine; ADOMET, S-adenosylmethionine; 2,4-DAB, 2,4-diaminobutyrate; A-2-C, azetidine-2-carboxylate; AIB, 2-amino-4-(isoxalin-5-one-2-yl)butyrate.

and synthesis of amino acids. Although we have not covered all of the "protein" amino acids in detail, it is clear that synthesis of some amino acids involves tightly regulated reactions, while for others there is little evidence of any control in their formation. Lys, his, thr, met, leu, val, ile and the aromatics are most tightly regulated; less rigid regulation is involved for arg, pro and cys, while for ala, ser, gly, glu, asp and their amides the supply of substrates may be the main regulatory factor. There is no obvious significance for this variation, although many of the latter group are greatly involved in transport and have a central role in a range of metabolic reactions; it is interesting that the group of more tightly regulated amino acids coincides with the group of amino acids essential in human diet.

In many species, large amounts of nitrogen reserve material is present and can support the growth of the seedling until well advanced; much of the discussion has been concerned with such seeds. However it should be emphasized that the growing embryo may not have an absolute dependence on stored organic nitrogen,[96a] and small seeded species may have an extremely limited supply of stored nitrogen.

While a number of metabolic events are characteristic of germination, the developing seedling is approaching autonomy and shows processes characteristic of the mature plant. The diversity of reactions present, the physiological importance of the event as well as the convenience as experimental material explains the popularity of germinating seeds as model systems.

ACKNOWLEDGMENTS

We are indebted to our colleagues both at Carleton and Rothamsted for many stimulating discussions on amino acid metabolism. In particular we would like to single out Drs. S. W. J. Bright, R. J. Ireland, B. J. Miflin and S. E. Rognes for their particular help in the preparation of this review.

REFERENCES

1. FOLKES BF, EW YEMM 1956 The amino acid content of the proteins of barley grains. Biochem J 62: 4-11
2. SODEK L, CM WILSON 1973 Metabolism of lysine and leucine derived from storage protein during germination of maize. Biochim Biophys Acta 304: 353-362
3. SOSULSKI FW, NW HOLT 1980 Amino acid composition and nitrogen-to-protein factors for grain legumes. Can J Plant Sci 60: 1327-1331
4. CHIBNALL AC 1939 Protein Metabolism in the Plant. Yale University Press, New Haven
5. STEWART CR, H BEEVERS 1967 Gluconeogenesis from amino acids in germinating castor bean endosperm and its role in transport to the embryo. Plant Physiol 42: 1587-1595
6. LARSON LA, H BEEVERS 1965 Amino acid metabolism in young pea seedlings. Plant Physiol 40: 424-432
7. BEEVERS L, WE SPLITTSTOESSER 1968 Protein and nucleic acid metabolism in germinating peas. J Exp Bot 19: 698-711
8. JOY KW, N NISHIKAWA 1979 Cautionary note: equilibration of label in amino acids by aminotransferases. Plant Physiol 63: 46s

9. COLLINS DM, AT WILSON 1975 Embryo and endosperm metabolism of barley seeds during early germination. J Exp Bot 26: 737-740
10. CHOU K-H, WE SPLITTSTOESSER 1972 Changes in amino acid content and the metabolism of γ-aminobutyrate in Cucurbita moschata seedlings. Physiol Plant 26: 110-114
11. KERN R, MJ CHRISPEELS 1978 Influence of the axis on the enzymes of protein and amide metabolism in the cotyledons of mung bean seedlings. Plant Physiol 62: 815-819
12. URQUHART AA, KW JOY 1981 Use of phloem exudate technique in the study of the amino acid transport in pea plants. Plant Physiol 68: 750-754
13. DURZAN DJ, V CHALUPA 1968 Free amino acids and soluble proteins in the embryo of jack pine as related to climate at the seed source. Can J Bot 46: 417-428
14. WINTER HC, GK POWELL, EE DECKER 1981 4-Methylene glutamine in peanut plants. Plant Physiol 68: 588-593
15. MAZELIS M 1980 Amino acid catabolism. In BJ Miflin, ed, Amino Acids and Derivatives, Biochemistry of Plants, Vol 5, Academic Press, New York pp 541-567
16. WIGHTMAN F, JC FOREST 1978 Properties of plant aminotransferases. Phytochemistry 17: 1455-1471
17. WILSON DG, KW KING, RH BURRIS 1954 Transamination reactions in plants. J Biol Chem 208: 863-874
18. FOREST JC, F WIGHTMAN 1972 Amino acid metabolism in plants II. Can J Bot 50: 538-542
19. FOREST JC, F WIGHTMAN 1971 Metabolism of amino acids in plants I. Can J Bot 49: 709-720
20. SANWAL BD, M LATA 1964 Enzymes of amino acid metabolism. In HF Linskens, BD Sanwal, MV Tracey, eds, Modern Methods of Plant Analysis, Springer, Berlin pp 290-360
21. KING J, WY WU 1971 Partial purification and kinetic properties of glutamic dehydrogenase from soybean cotyledons. Phytochemistry 10: 915-928
22. THURMAN DA, C PALIN, MV LAYCOCK 1965 Isoenzymatic nature of L-glutamic dehydrogenase of higher plants. Nature 207: 193-194
23. CHOU K-H, WE SPLITTSTOESSER 1972 Glutamate dehydrogenase from pumpkin cotyledons. Plant Physiol 49: 550-554

24. YUE SB 1969 Isoenzymes of glutamate dehydrogenase in plants. Plant Physiol 44: 453-457
25. STEWART GR, AF MANN, PA FENTEM 1980 Enzymes of glutamate formation: glutamate dehydrogenase, glutamine synthetase, and glutamate synthase. In BJ Miflin, ed, Amino Acids and Derivatives, Biochemistry of Plants, Vol 5, Academic Press, New York pp 271-327
26. MIFLIN BJ, PJ LEA 1982 Ammonia assimilation and amino acid metabolism. In D Boulter and B Parthier, eds, Encyclopedia of Plant Physiology, New Series, Vol 14A, Springer Verlag, Berlin pp 5-64
27. MIFLIN BJ, PJ LEA 1980 Ammonia assimilation. In BJ Miflin, ed, Amino Acids and Derivatives, Biochemistry of Plants, Vol 5, Academic Press, New York pp 169-202
28. VARNER JE, GC WEBSTER 1955 Studies on the enzymatic synthesis of glutamine. Plant Physiol 30: 393-402
29. WEBSTER GC, JE VARNER 1955 Aspartate metabolism and asparagine synthesis in plant systems. J Biol Chem 215: 91-99
30. ROGNES SE 1975 Glutamine-dependent asparagine synthetase from *Lupinus luteus*. Phytochemistry 14: 1975-1982
31. LEA PJ, L FOWDEN 1975 The purification and properties of glutamine-dependent asparagine synthetase isolated from *Lupinus albus*. Proc Roy Soc Lond B 192: 13-26
32. STREETER J 1973 *In vivo* and *in vitro* studies on asparagine biosynthesis in soybean seedlings. Arch Biochem Biophys 157: 613-624
33. LEA PJ, BJ MIFLIN 1980 Transport and metabolism of asparagine and other nitrogen compounds within the plants. In BJ Miflin, ed, Amino Acids and Derivatives, Biochemistry of Plants, Vol 5, Academic Press, New York pp 569-607
34. MATOH T, E TAKAHASHI 1982 Changes in the activities of ferredoxin- and NADH-glutamate synthase during seedling development of peas. Planta 154: 289-294
35. STREETER J 1977 Asparaginase and asparagine transaminase in soybean leaves. Plant Physiol 60: 235-239
36. SODEK L, PJ LEA, BJ MIFLIN 1980 Distribution and properties of a potassium-dependent asparaginase isolated from developing seeds of *Pisum sativum* and other plants. Plant Physiol 65: 22-26
37. ISLAND RJ, KW JOY 1981 Two routes for asparagine metabolism in *Pisum sativum* L. Planta 151: 289-292

38. BAUER A, KW JOY, AA URQUHART 1977 Amino acid metabolism of pea leaves - labeling studies on utilization of amides. Plant Physiol 59: 920-924
39. LLOYD NDH, KW JOY 1978 2-Hydroxysuccinamic acid: a product of asparagine metabolism in plants. Biochem Biophys Res Comm 81: 186-192
40. KOLLOFFEL C, HD VAN DIJKE 1975 Mitochondrial arginase activity from cotyledons of developing and germinating seeds of _Vicia faba_. Plant Physiol 55: 507-510
41. JONES VM, D BOULTER 1968 Arginine metabolism in germinating seeds of some members of the Leguminosae. New Phytol 67: 925-934
42. SPLITTSTOESSER WE 1969 The appearance of arginine and arginase in pumpkin cotyledons, characterization of arginase. Phytochemistry 8: 753-758
43. SEHGAL PP, AW NAYLOR 1966 Ontonogenic study of urease in jack beans, _Canavalia ensiformis_ (L) DC. Bot Gaz 127: 27-34
44. SUZUKI Y 1952 Increase in level of arginase and urease in germinating soybean. Kagaku, Tokyo 22: 264-265
45. ROSENTHAL GA 1970 Canavanine utilization in the developing plant. Plant Physiol 46: 273-276
46. WHITESIDE JA, DA THURMAN 1971 The degradation of canavanine by jack bean cotyledons. Planta 98: 279-284
47. TROMP J, JC OVAA 1979 The transport of arginine and asparagine in apple trees. Physiol Plant 45: 23-28
48. BIDWELL RGS, DJ DURZAN 1975 Some recent aspects of nitrogen metabolism. _In_ PJ Davies, ed, Historical and Current Aspects of Plant Physiology, Cornell University Press, New York pp 152-225
49. SMITH TA 1977 Recent advances in the biochemistry of plant amines. Progr Phytochem 4: 27-82
50. SHARGOOL PD, EA COSSINS 1968 Further studies of L-arginine biosynthesis in germinating pea seeds. Can J Biochem 46: 393-399
51. O'NEAL TD, AW NAYLOR 1969 Partial purification and properties of carbamyl phosphate synthetase of Alaska pea. Biochem J 113: 271-279
52. KLECZKOWSKI K, P COHEN 1964 Purification of ornithine transcarbamylase from pea seedlings. Arch Biochem Biophys 107: 271-278
53. EID S, Y WALY, AT ABDELAL 1974 Separation and properties of two ornithine carbamoyl-transferases from _Pisum sativum_ seedlings. Phytochemistry 13: 99-102

54. SHARGOOL PD 1971 Purification of arginosuccinate synthetase from cotyledons of germinating peas. Phytochemistry 10: 2029-2032
55. SHARGOOL PD, EA COSSINS 1968 Isolation and some properties of arginosuccinate lyase from a higher plant. Can J Biochem 46: 393-399
56. TAYLOR AA, GR STEWART 1981 Tissue and subcellular localization of enzymes of arginine metabolism in *Pisum sativum*. Biochem Biophys Res Comm 101: 1281-1289
57. ROSENTHAL GA 1982 L-Canavanine metabolism in jack bean, *Canavalia ensiformis* DC. Plant Physiol 69: 1066-1069
58. MIFLIN BJ, PR SHEWRY 1979 The biology and biochemistry of cereal seed prolamins. *In* Seed Protein Improvement in Cereals and Grain Legumes, Vol 1, IAEA, Vienna pp 137-158
59. WANG D 1969 Metabolism of amino acids and amines in germinating seeds. Contrib Boyce Thompson Inst 24: 109-115
60. JONES M, JS PIERCE 1967 The role of proline in the amino acid metabolism of germinating barley. J. Inst Brew 73: 577-583
61. THOMPSON JF 1980 Arginine synthesis, proline synthesis and related processes. *In* BJ Miflin, ed, Amino Acids and Derivatives, Biochemistry of Plants, Vol 5, Academic Press, New York pp 375-402
62. OAKS A, IJ MITCHELL, RA BARNARD, FT JOHNSON 1970 The regulation of proline biosynthesis in maize roots. Can J Bot 48: 2249-2258
63. BOGGESS SF, D ASPINALL, LF PALEG 1976 The significance of end-product inhibition of proline biosynthesis and of compartmentation in relation to stress induced proline accumulation. Aust J Pl Physiol 3: 513-525
64. MORRIS CJ, JF THOMPSON, CM JOHNSON 1969 Metabolism of glutamic acid and N-acetyl glutamic acid in leaf discs and cell free extracts of higher plants. Plant Physiol 44: 1023-1026
65. BOGGESS SF, CR STEWART, D ASPINALL, LG PALEG 1976 Effects of water stress on proline synthesis from radioactive precursors. Plant Physiol 58: 398-401
66. RENA AB, WE SPLITTSTOESSER 1975 Proline dehydrogenase and pyrroline-5-carboxylate reductase from pumpkin seedlings. Phytochemistry 14: 657-661
67. VANSUYT G, JC VALLEE, J PREVOST 1979 Pyrroline-5-carboxylate reductase and proline dehydrogenase in

Nicotiana tabacum as a function of its development. Physiol Veg 17: 95-105

68. MESTICHELLI LJJ, RN GUPTA, ID SPENSER 1979 The biosynthetic route from ornithine to proline. J Biol Chem 254: 640-647
69. MAZELIS M, L FOWDEN 1969 Conversion of ornithine into proline by enzymes from germinating peanut cotyledons. Phytochemistry 8: 801-810
70. SPLITTSTOESSER WE, L FOWDEN 1973 Ornithine transaminase from *Cucurbita maxima* cotyledons. Phytochemistry 12: 1565-1568
71. LU TS, M MAZELIS 1975 L-Ornithine: 2-oxo acid aminotransferase from squash (*Cucurbita pepo* L.) cotyledons. Plant Physiol 55: 502-506
72. HASSE K, OT RATYCH, J SALKINOW 1967 Transamination and decarboxylation of ornithine in higher plants. Hoppe-Selyer's Z Physiol Chem 348: 843-851
73. SENEVIRATNE AS, L FOWDEN 1968 Diamino acid metabolism in plants with special reference to α,ε-diaminopimelic acid. Phytochemistry 7: 1047-1056
74. MEISTER A, AN RADHAKRISHNAN, SD BUCKLEY 1957 Enzymatic synthesis of L-pipecolic acid and L-proline. J Biol Chem 229: 789-800
75. MACHOLAN L, P ZOBAL, J HEKELOVA 1965 Activity, utilization and metabolism of 5-amino-2-oxovaleric acid in pea seedlings and bakers yeast. Hoppe-Selyer's Z Physiol Chem 349: 97-106
76. GAVAZZI G, M RACCHI, C TONELLI 1975 A mutation causing proline requirement in *Zea mays*. Theor Appl Genet 46: 339-346
77. BERTANI A, C TONELLI, G GAVAZZI 1980 Determination of Δ^1-pyrolline-5-carboxylate reductase in proline requiring mutants of *Zea mays* L. Maydica 25: 17-24
78. BRIGHT SWJ, PJ LEA, JSH KUEH, C WOODCOCK, DW HOLLOMAN, GC SCOTT 1982 Proline content does not influence pest and disease susceptibility of barley. Nature 295: 592-593
79. MILLS WR, PJ LEA, BJ MIFLIN 1980 Photosynthetic formation of the aspartate family of amino acids in isolated chloroplasts. Plant Physiol 65: 1166-1172
80. WALLSGROVE RM, M MAZELIS 1980 The enzymology of lysine biosynthesis in higher plants. FEBS Lett 116: 189-192
81. SAINIS JK, RG MAYNE, RM WALLSGROVE, PJ LEA, BJ MIFLIN 1981 Localization and characterization of homoserine

dehydrogenase isolated from barley and pea leaves. Planta 152: 491-496

82. WALLSGROVE RM, PJ LEA, BJ MIFLIN 1983 The intracellular localization of the enzymes of threonine and methionine biosynthesis in green leaves. Plant Physiol (in press)
83. WALLSGROVE RM, M MAZELIS 1981 The enzymology of lysine biosynthesis in higher plants; partial purification and characterization of spinach leaf dihydropicolinate synthase. Phytochemistry 20: 2651-2655
84. SHEWRY PR, BJ MIFLIN 1977 Properties and regulation of aspartate kinase from barley seedlings. Plant Physiol 59: 69-73
85. AARNES H 1977 A lysine-sensitive aspartokinase and two molecular forms of homoserine dehydrogenase from barley seedlings. Plant Sci Lett 9: 137-145
86. BRIGHT SWJ, PR SHEWRY, BJ MIFLIN 1978 Aspartate kinase and the synthesis of aspartate derived amino acids in wheat. Planta 139: 119-125
87. LEA PJ, WR MILLS, BJ MIFLIN 1979 The isolation of a lysine-sensitive aspartate kinase from pea leaves and its involvement in homoserine biosynthesis in isolated chloroplasts. FEBS Lett 98: 165-168
88. ROGNES SE, PJ LEA, BJ MIFLIN 1980 _S_-adenosylmethionine, a novel regulator of aspartate kinase. Nature 287: 375-359
89. MADISON JT, JF THOMPSON 1976 Threonine synthetase from higher plants: stimulation by _S_-adenosylmethionine and inhibition by cysteine. Biochem Biophys Res Comm 71: 684-691
90. AARNES H 1978 Regulation of threonine and biosynthesis in barley seedlings. Planta 140: 185-192
91. AARNES H, SE ROGNES 1974 Threonine-sensitive aspartate kinase and homoserine dehydrogenase from _Pisum sativum_. Phytochemistry 13: 2717-2724
92. BRYAN JK, NR LOCHNER 1981 The effects of plant age and extraction conditions on the properties of homoserine dehydrogenase isolated from maize seedlings. Plant Physiol 68: 1395-1399
93. MATTHEWS BF, JM WIDHOLM 1979 Regulation of homoserine dehydrogenase in developing organs of soybean seedlings. Phytochemistry 18: 395-400
94. DI MARCO G, S GREGO 1975 Homoserine dehydrogenase in _Pisum sativum_ and _Ricinus communis_. Phytochemistry 14: 943-947

95. THOMPSON GA, AH DATKO, SH MUDD, J GIOVANELLI 1982 Methionine biosynthesis in Lemna. Plant Physiol 69: 1077-1083

96a. JOY KW, BF FOLKES 1965 The uptake of amino acids and their incorporation into the proteins of excised barley embryos. J Exp Bot 16: 646-666

96b. BRIGHT SWJ, EA WOOD, BJ MIFLIN 1978 The effect of aspartate-derived amino acids (lysine, threonine, methionine) on the growth of excised embryos of wheat and barley. Planta 139: 113-117

97. BRIGHT SWJ, PJ LEA, BJ MIFLIN 1980 The regulation of methionine biosynthesis and metabolism in plants and bacteria. In K Elliott, J Wheland, eds, Ciba Foundation Symp 72, Sulphur in Biology, Exerpta Medica, Amsterdam pp 101-114

98. BRIGHT SWJ, SE ROGNES, BJ MIFLIN 1982 Threonine accumulation in the seeds of a barley mutant with altered aspartate kinase. Biochem Genet 20: 229-243

99. BRIGHT SWJ, JSH KUEH, J FRANKLIN, SE ROGNES, BJ MIFLIN 1982 Two genes for threonine accumulation in barley seeds. Nature 299: 278-279

100. ROGNES SE, SWJ BRIGHT, BJ MIFLIN 1982 Feedback insensitive aspartate kinase isoenzymes in barley mutants resistant to lysine plus threonine. Planta (in press)

101. SAKANO H 1979 Derepression and repression of lysine-sensitive aspartokinase during in vitro culture of carrot root tissue. Plant Physiol 63: 583-585

102. BRYAN PA, RD CRAWLEY, CE BRUNNER, JK BRYAN 1970 Isolation and characterization of a lysine sensitive aspartokinase from a multicellular plant. Biochem Biophys Res Comm 41: 1211-1217

103. MATTHEWS BF, AW GURMAN, JK BRYAN 1975 Changes in enzyme regulation during growth of maize. I. Progressive desensitization of homoserine dehydrogenase during seedling growth. Plant Physiol 55: 991-998

104. DICAMELLI CA, JK BRYAN 1975 Changes in enzyme regulation during growth of maize. II. Relationships amongst multiple molecular forms of homoserine dehydrogenase. Plant Physiol 55: 999-1005

105. BRYAN JK, EA LISSIK, BF MATTHEWS 1977 Changes in enzyme regulation during growth of maize. III. Intracellular location of homoserine dehydrogenase in chloroplasts. Plant Physiol 59: 673-679

106. BRYAN JK, NR LOCHNER 1981 Quantitative estimates of the distribution of homoserine dehydrogenase isoenzymes in maize tissues. Plant Physiol 68: 1400-1405

107. AARNES H 1980 Biosynthesis of the thioether cystathionine in barley seedlings. Plant Sci Lett 19: 81-89

108. MAZELIS M, FR WHATLEY, J WHATLEY 1977 The enzymology of lysine biosynthesis in higher plants. FEBS Lett 84: 236-240

109. LAWRENCE JM 1973 Homoserine in seedlings of the tribe Vicieae of the Leguminosae. Phytochemistry 12: 2207-2209

110. VIRTANEN AI, A BERG, S KARI 1953 Formation of homoserine in germinating pea seeds. Acta Chem Scand 7: 1423-1424

111. LAWRENCE JM, DR GRANT 1963 Nitrogen mobilization in pea seedlings. II. Free amino acids Plant Physiol 38: 561-566

112. LARSON LA, H BEEVERS 1965 Amino acid metabolism in young pea seedlings. Plant Physiol 40: 424-432

113. VAN EGERAAT AWSM 1975 Changes in free ninhydrin-positive compounds of young pea plants as affected by different nutritional and environmental conditions. Plant Soil 42: 15-36

114. NAYLOR AW, R ROBSON, NE TOLBERT 1958 Aspartic-C^{14} acid metabolism in leaves, roots and stems. Physiol Plant 11: 537-547

115. MITCHELL DJ, RGS BIDWELL 1970 Compartments of organic acids in the synthesis of asparagine and homoserine in pea roots. Can J Bot 48: 2001-2007

116. MILLS WR, KW JOY 1980 A rapid method for isolation of purified physiologically active chloroplasts and its use to study the intracellular distribution of amino acids in pea leaves. Planta 148: 75-83

117. GRANT DR, E VOELKERT 1971 The formation of homoserine from methionine in germinating peas. Can J Biochem 49: 795-798

118. MATTHEWS BF, JM WIDHOLM 1979 Enzyme expression in soyabean cotyledons and cell suspension cultures. Can J Bot 57: 299-304

119. GREGO S, D TRICOLL, G DIMARCO 1980 Comparisons of homoserine dehydrogenase from different plant sources. Phytochemistry 19: 1619-1623

120. THOEN A, SE ROGNES, H AARNES 1978 Biosynthesis of threonine from homoserine in pea seedlings. I. Homoserine kinase. Plant Sci Lett 13: 103-112

121. MITCHELL DJ, RGS BIDWELL 1970 Synthesis and metabolism of homoserine in developing pea seedlings. Can J Bot 48: 2037-2042
122. NIGRAM SN, C RESSLER 1966 Biosynthesis of 2,4-diaminobutyric acid from L-[^{3}H]-homoserine and DL-[^{14}C]-aspartic acid in *Lathyrus sylvestris* L. Biochemistry 5: 3426-3431
123. GROBBELAAR N, FC STEWARD 1969 The isolation of amino acids from *Pisum sativum*; identification of L(-)-homoserine and L(+)-*O*-acetylhomoserine and certain effects of the environment upon their formation. Phytochemistry 8: 553-559
124. PRZBYLSKA J, J PAWELKIEWICZ 1965 *O*-oxalyl homoserine, a new homoserine derivative in young pods of *Lathyrus sativus*. Bull Acad Pol Sci Ser Biol 13: 327-329
125. SUNG M-L, L FOWDEN 1971 Imino acid biosynthesis in *Delonix regia*. Phytochemistry 10: 1523-1528
126. LAMBEIN F, Y-H KUO, R VAN PARIJS 1976 Isoxazolin-5-ones. Chemistry and biology of a new class of plant products. Heterocycles 4: 567-593

Chapter Six

STARCH-LIPID COMPLEXES AND OTHER NON-STARCH COMPONENTS OF STARCH GRANULES IN CEREAL GRAINS

T. GALLIARD

RHM Research Ltd.
Lord Rank Research Centre
Lincoln Road
High Wycombe, Buckinghamshire
HP12 3QR, United Kingdom

INTRODUCTION

Starch represents one of the major organic compounds in Nature and is the principal reserve polysaccharide of plants. It is a major component of the diet of man and many animals and starch is becoming an increasingly important, readily renewable resource as a raw material for the chemical and food industries.

Despite considerable efforts by chemists and biochemists over the past 40 years, starch science and technology is largely empirical because we cannot yet explain many of the functional (physical) and nutritional properties of starch in terms of its chemistry and biochemistry.

The commercially important sources of isolated starch in Europe and North America are cereal grains (maize, wheat) and root crops (potato). All starches exist naturally essentially as insoluble granules and their uses as foods or raw materials depend on methods to render the main polysaccharide α-glucan components accessible for reaction by either chemical or enzymic processes.

To compare the physical properties of starches, a standard approach used by the starch technologist is to measure viscosities of starch suspensions in a temperature-time profile. Fig. 1 shows major differences in the rheological properties of starch from three common sources, viz. potato, maize and wheat. Gross chemical analyses of these starches show no major differences and even the proportions of the amylose and amylopectin fractions in these starches are very similar (all containing 25-30% amylose, 70-75% amylopectin, see also Table 4, p. 120).[1,2] As the major differences in physical properties are not explicable in terms of gross composition of the main components, other explanations have been sought. Many laboratories have

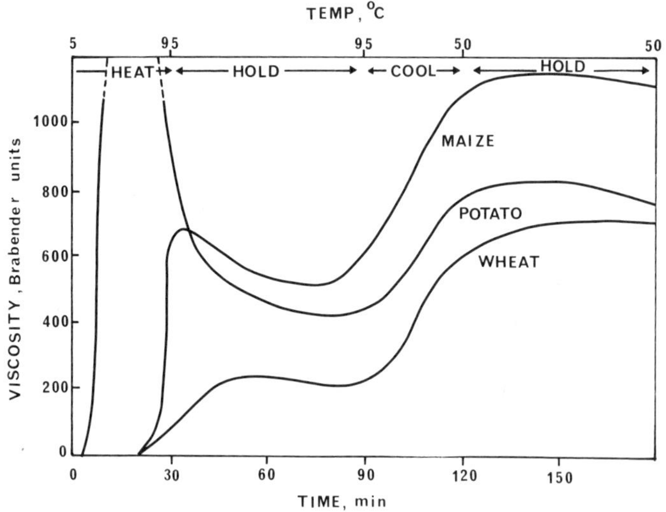

Fig. 1. Comparison of viscosity characteristics of pastes of commercial starch preparations from potato, maize and wheat. Starch preparations (10% in water) in a Brabender amylograph were heated to 95°C, held at this temperature for 60 min then cooled to 50°C. Viscosity (in arbitrary 'Brabender units') was recorded continuously. (By courtesy of Dr. D. Howling, of C.P.C. (U.K.) Ltd.)

investigated polymer fine structure, chain length, branching, spatial arrangements, crystalline or amorphous character, etc. of the polymer networks within starch granules and undoubtedly some physical differences between starches can be attributed to alignment of polymer chains, accessibility of water, packing densities, etc.[1,2] However, within the area of cereal grain starches, differences still exist that cannot be explained in these terms. We have recently become interested in the possible roles played by relatively minor components of starch granules from cereals. The following discussion will review currently available information from several laboratories actively interested in the roles of non-starch components of starch granules.

MINOR COMPONENTS OF CEREAL STARCHES

Components other than α-glucan polymers in starch granules can be considered either as <u>surface</u> materials on the granules or as <u>internal</u> components within the granule matrix.

Surface Components

Materials on the surface of granules would not appear as major components in a total analysis of starch but their presence could be very significant in determining the interaction of starch granules with their environment. Cereal starch granules range from 1 to 40 μm in diameter. A simple calculation shows that for a spherical particle of radius 10 μm, a surface coating of 10 nm thickness (equivalent to a biomolecular lipo-protein membrane) would represent only about 0.3% of the granule volume, i.e. within the range of experimental error for conventional gross analyses. Minimal values for minor components of carefully washed starches are within this range (Table 1).

Some surface materials on isolated starch granules may of course be artefacts of preparation and represent contaminants but, conversely, true surface material within the grain may be removed during isolation of starch. This point is particularly relevant if, as has been suggested,[3-5] starch and protein form a continuous matrix in mature cereal grains and that the extent of cohesion between starch granules and protein affects the physical properties of grain.

Table 1. Residual minor components of well-washed starch granules from wheat.

Starch preparation	Lipid	Protein		'Pentosan'
	as fatty acid methyl ester	Kjeldahl N x 5.7	amino acid analysis	as furfural after HCL hydrolysis
	% dry weight of starch			
Mixed large (lenticular) and small granules; from flour	0.40-0.45	0.2-0.3		0.2-0.3
Mixed large and small granules; from grain (after protease treatment)	0.3-0.4	0.15		0.1
Large (lenticular) granules only; from flour			0.1	0.2

Data from P. Bowler et al (unpublished) and Lowy et al.[11]

Unlike leaf or potato tuber tissues in which starch granules are present in fully hydrated cells and in which amyloplast membranes can be observed, mature cereal grains do not contain intact amyloplasts and the membranes within which the starch granules developed cannot be isolated. Amyloplast integrity is lost during later stages of ripening when loss of intracellular organization occurs. However, one might expect to see evidence of amyloplast membrane components (or their degradation products) associated with starch granules. In fact, the non-starch lipids of cereal endosperm (i.e. the lipids not within the starch granules) are relatively rich in glycolipids which are similar to, and may be derived from, the type of lipids found in amyloplast membrane in other tissues.[6]

Morrison's group has made detailed studies on the lipids associated with starch granules from cereals and has clearly distinguished the internal starch lipids, and the surface lipids on the basis of solvent extractability.[6] Essentially, the internal lipids are only extracted by hot, polar solvents that disrupt the granule structure; thus lipids extracted rapidly by polar solvents at ambient temperature may be considered to be associated with the surface of granules. Table 2 gives results for major lipids of wheat and maize starch.

Table 2. Ranges for percentage composition of starch internal lipid and endosperm lipid in wheat and maize.

Lipid	Percent of total lipid			
	Maize[a]		Wheat[a]	
	Internal starch lipid	Endosperm lipid[b]	Internal starch lipid	Endosperm lipid[b]
Triglycerides	1-2	36-54	0.3-0.5	17-34
Free fatty acids	55-61	21-52	2.9-3.5	3-7
Glycolipid	3-9	2-7	1-7	21-38
N-Acylphospholipid	0	trace	0	12-21
Diacylphospholipid	0	3	0-1	4.8
Lysophospholipid	26-34	5	88-94	6-13
Total lipid: (% endosperm dry weight)	0.6-0.8	0.8-1.9	0.8-1.2	0.9-1.1

[a] Range for several varieties.

[b] Includes starch surface lipid but excludes starch internal lipid.

Adapted from Morrison and Milligan (1982).[8]

The high proportion of glycolipids in endosperm is typical of lipids derived from plastid membranes (chloroplasts, amyloplasts, etc.) as illustrated by some recent analyses of amyloplast membranes from immature potato tubers[7] (see Table 3). Endosperm lipid of wheat also contains substantial amounts of neutral lipids, including triglycerides, but these are derived from spherosomes rather than membranes.[10]

Well-washed starch from wheat retains some protein (Table 1), most of which is extractable only after disruption of the granule in hot dissociating conditions. Recent work by Lowy et al[11] in our laboratories has found an unusual protein in wheat starch preparations. This protein is also presumed to be associated with the surface of wheat starch granules, since it is readily extracted with dilute salt under mild conditions that cause no disruption of granules. The 'surface' protein is mainly represented by a single species of protein, molecular weight ≅ 30,000 with a high ρI (>10). The protein is not found in significant amounts in the endosperm proteins and is therefore not a result of non-specific contamination of starch granules with endosperm material during starch isolation. Whether the protein is associated with starch granules in intact tissue is not yet known and current studies with immunological techniques should answer this point. It is known that amylases and other proteins can bind to starch granules but the surface

Table 3. Comparison of phospho- and glyco-lipid proportions in wheat endosperm and potato amyloplast membranes.

	Wheat endosperm[9]		Potato amyloplast[6]
Membrane lipids	Internal starch	Endosperm (including starch surface)	
	%	%	%
Phospholipids	94	38	30
Glycolipids	6	62	70

Data adapted from: Hargin and Morrison (1980)[9] and Fishwick and Wright (1980)[6]

protein described here has neither the activity nor size and charge characteristics of α-amylase, β-amylase or the α-amylase inhibitors of wheat. This surface protein is not found in starch prepared from partially germinated wheat.

In germinating seeds, the enzymic breakdown of starch molecules depends upon access of enzymes to substrates. The outer surfaces of starch granules may present a barrier to amylolytic enzymes, possibly due to non-starch material on the surface or due to the physical arrangement of the amylose and amylopectin chains near to the surfaces of granules. Fig. 2 shows scanning electron microscopic images of the surface and interior of lenticular starch granules from wheat germinated 5 days at 20°C. The main points to note are a) the preferential digestion around the equatorial groove; b) the greater digestion of the interior than the surface of the granule; c) the attack at relatively few sites on the surface and the preferential further attack at these sites; d) the layers of higher and lower susceptibility to enzyme attack. The apparent resistance of much of the surface to enzyme attack is clearly illustrated.

There is little information on other possible surface components of starch granules. Non-starch polysaccharides, associated with isolated starch granules (Table 1), are almost certainly present in contaminating cell-wall materials: careful purification methods reduce the levels of non-starch polysaccharides to the limit of detection (P. Bowler, unpublished). Examination by micro-electrophoresis of starch granules from various sources has demonstrated the differences in surface charges and the effects of selective removal of surface components (lipids, proteins, cations) on zeta potentials of starch granules. The fact that wheat starch granule dispersions are destabilized at the pH value giving zero zeta potential,[13] indicates that starch granule dispersions are charge-stabilized.

Internal Components

Proteins. In addition to surface protein, described above, wheat starch granules contain proteins (approximately 0.1% by weight of starch) that are not released by dissociating agents (e.g. sodium dodecyl sulphate) until the granules have been gelatinized by heating.[11] The disruptive conditions required to release these proteins indicate that they are buried within the matrix of the granules. These conditions

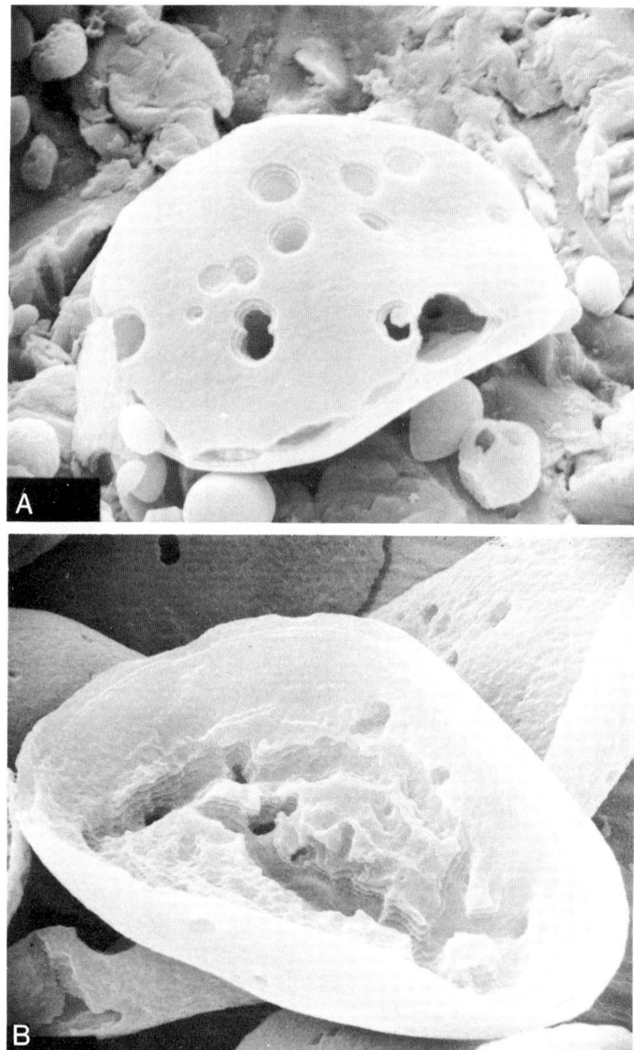

Fig. 2. Scanning electron micrographs of starch granules from wheat grain germinated for 5 days at 20°C. Partly digested granules (as in Fig. 2A) readily fragment along the equatorial groove to form half-granules as in Fig. 2B. (From J. G. Sargeant;[12] reproduced with permission).

also make it difficult to assess the proteins for native structure, enzyme activity, etc. The subunits obtained on SDS-polyacrylamide gel electrophoresis are of higher molecular weight than the surface protein described above.

It is conceivable that the internal proteins may represent residual material from lipo-protein membranes of the original amyloplasts or of membrane-bound starch synthesizing systems employed during development. The presence of protein/enzymes associated with starch biosynthesis in developing granules and amyloplasts is reviewed by Badenhuizen.[14] Lipid, which may also be derived from lipo-protein membranes, is present in the granules as detailed below.

Lipids. The lipid components of starch granules have attracted considerable attention recently for several reasons: 1) major differences in content and composition of internal lipids of starch granules from different species and cultivars have been confirmed; 2) lipids have been shown to affect the viscosity characterisitics of starch in systems relevant to a range of industrial processes and 3) possible roles for starch lipids in both biosynthesis and degradation of starch have been investigated, but relatively little is yet known of the mechanisms and control of starch accumulation within granules or the subsequent breakdown of granule structures.

The presence in some starches of lipid materials that form complexes with amylose, has been known for many years[14,15] and early work has been reviewed by Acker[16] in whose laboratory in Germany, detailed studies of starch-lipid complexes have been made more recently. Other recent studies, particularly from Morrison's group in Glasgow, have refined the methods for extraction and analysis of starch lipids to a point now where a much clearer picture of lipid-starch relationships is emerging.

Although complex formation between amylose and certain lipids is a general phenomenon and occurs with amylose from both cereal and non-cereal sources, only the cereal starches have so far been shown to contain endogenous internal starch lipids i.e. lipids occur naturally within the starch granules. This is illustrated in Table 4 which gives data for the lipid content of some cereal starches (wheat, maize, rice) compared with other, non-cereal starch preparations. In Table 4 lipid

Table 4. Internal lipid and amylose content of some starches.

Source of starch	Lipid[a]	Amylose: Apparent[b]	Amylose: Total[b]
	mg/100 g starch	% of total starch	
Rice grain	662	15.2	21.9
Maize grain	566	21.4	27.5
Wheat grain	426	20.4	27.2
Potato tuber	16	26.0	25.9
Vicia faba cotyledon	12	42.3	42.1
Mung bean cotyledon	69	40.1	40.0
Parsnip root	15	24.2	24.4

[a] Expressed as fatty acid methyl esters on dry weight basis.

[b] Amylose measured by iodine binding before (apparent) and after (total) removal of internal lipid.

Data from Morrison and Laignelet (1983).[17]

contents are expressed in terms of fatty acid methyl ester equivalents; calculated as total lipid, the cereal starches contain around 1% lipid on a dry weight basis. It can also be seen that those (cereal) starches containing internal lipid also give major discrepancies in amylose content measured by I_2-binding if lipid is not removed prior to amylose determination. My colleague Jim Sargeant has recently developed an alternative analytical (based on enzymic debranching of amylopectin in DMSO solution) method for amylose/amylopectin ratio measurement in cereal starches[18] and this has been used to standardize a more rapid colorimetric assay[17] for amylose determinations in starch-lipid complexes.

Detailed analyses of the internal lipids of cereal starches have confirmed several novel features. Table 2 presents analyses of internal and endosperm lipids of starch from wheat and maize. Firstly, it should be noted that the internal lipid composition is relatively simple; around 90% of wheat starch internal lipid is lysophospholipid (of which lysophosphatidylcholine is the major component); monoacylglycolipids and free fatty acids constitute the remaining few percent of lipid. In maize, free fatty acids represent the major class of internal starch lipid (around 60%) with lysophospholipid and glycolipid making up the remaining lipid.

Secondly, it is significant that the internal lipids associated with starch are exclusively monoacyl lipids, (see Fig. 3), i.e. they have only one fatty acyl chain per molecule. Such lipids are not common in living tissue where membrane systems will normally only contain diacyl lipids; lysophospholipids are extremely cytotoxic (the lyso-prefix is derived from their ability to cause lysis of erythrocytes and other membrane systems) and free fatty acids also react with membranes and proteins. The presence of monoacyl polar lipids and free fatty acids in tissues is often a sign of loss of cellular integrity. The high content of lysophospholipids in cereal starches is illustrated in Table 5 which summarizes recent results from two laboratories. Ito et al[19] listed a range of lipids in rice starch but recent studies cast doubt on the classification of all these as internal lipids of starch (W. R. Morrison, personal communication).

The fatty acid component of the lysophospholipid is a mixture, mainly of the two major fatty acids of cereal grains, i.e. palmitic acid ($C_{16:0}$, saturated), linoleic acid ($C_{18:2}$, diunsaturated).

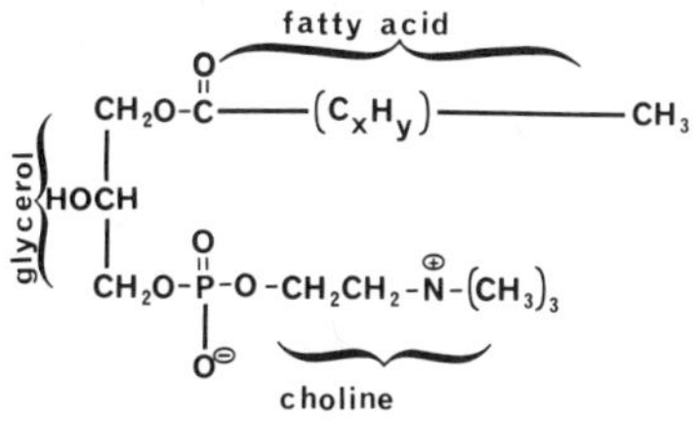

Fig. 3. Structure of lysophosphatidylcholine, the major lipid component of the amylose-lipid complex in wheat starch. Note: phospholipids as normally found in tissues are diacylphospholipids containing a second fatty acyl group on the 2-position of the glycerol moiety.

In maize starches there is a close relationship between the lipid content and the proportion of amylose in the starch.[20] This is well illustrated with analyses of starches from maize mutants having relatively high (amylomaize), normal and very low (waxy maize) proportions of amylose (Table 6). It is clear that in maize, starch free fatty acid content and lysophospholipid content are both related to the amount of amylose in the starch. However this relationship between amylose and lipid content is less evident in rice starches (W. R. Morrison, personal communication).

It is generally assumed that the internal starch lipid _in situ_ is present as a complex with amylose and that the fatty acyl chain lies within helical regions of the amylose

Table 5. Major lipids of some cereal starches.

Cereal	Percent of total internal starch lipid		Reference
	Lysophospholipid	Free fatty acid	
Wheat	88-94	3-4	8
Maize	26-34	55-61	8
Rice	60-83	27-40	a
Barley	72	4	16
Oat	64	8	16
Rye	52	2	16

[a] W. R. Morrison (personal communication).

Table 6. Internal starch lipids of maize varieties. (Range of values obtained for amylose and lipid content.)

	Amylose content of starch		
	Low	Normal	High
No. of samples	3	20	5
Amylose content (%)	0.2-1.5	26.6-31.1	40.0-46.9
Lipid content ($mg \cdot 100\ g^{-1}$)			
Free fatty acids	3-13	380-546	581-681
Lysophospholipids	2-13	184-347	396-486

Data from Morrison and Milligan (1982).[8]

polymer as indicated by the following representation (Fig. 4). In the conventional structure for the amylose helix the hydrophylic (OH) groups are usually depicted on the outside of the helix, leaving a hydrophobic core the diameter of which could accommodate one, but not more, fatty acid chains. Amylose-lipid complexes are readily prepared and their structures have been established mainly from X-ray studies.[15,16] However, naturally-occurring cereal starches do not give characteristic X-ray 'V patterns' of amylose-lipid complexes, possibly because insufficient regular order is present to generate a diffraction pattern. Internal lipids of cereal starch granules interfere with I_2-binding to cereal starches (Table 4) and the normally reactive linoleic acid (18:2) is resistant to oxidation (1 month in O_2 at 70°C) and chlorination when present in the lysophospholipid-amylose complex in wheat starch.[21] Nevertheless, Morrison[8] concludes that available evidence does not preclude an association between lipid and amylose chains other than the lipid-in-helix model.

Assuming a helical structure for the amylose-lipid complex, the polar group of complexed lysophospholipid is presumably at least partly exposed outside the helical segments since, as shown by Acker and colleagues,[16] amylose-bound lysophosphatidylcholine is attacked by phospholipase D, which acts as the choline-phosphate ester group but not by acyl hydrolases, which act at the fatty acyl ester bond as shown in Fig. 5.

What proportion of the amylose polymer chains in cereal starches is present as a lipid complex? Examination of

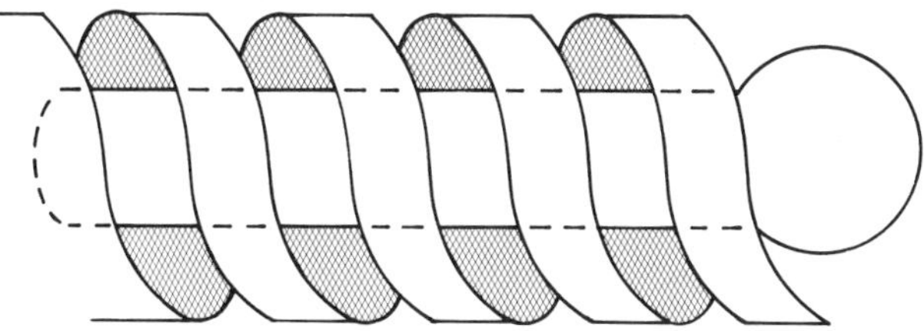

Fig. 4. Diagrammatic representation of a monoacyl lipid within a helical section of an amylose 1,4 α-glucan chain. The hydrophobic, paraffinic portion of the lipid molecule (but not the polar head group) is shown within the amylose helix.

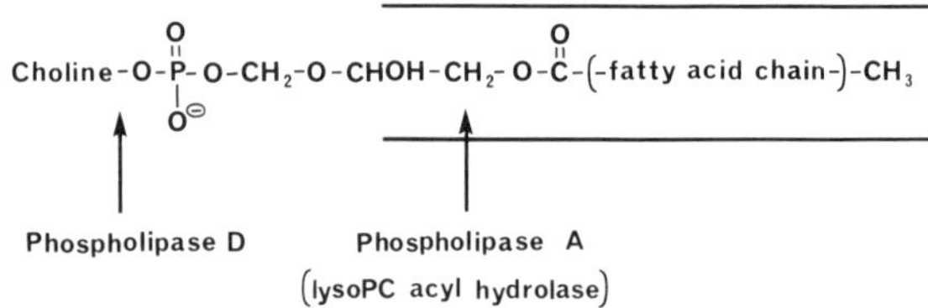

Fig. 5. Sites of action of phospholipases A and D on lysophosphatidylcholine. The horizontal lines indicate the portion of the lipid molecule assumed to be complexed with amylose.

Table 4 indicates that the presence of lipid decreases the I_2-binding capacity by 20 to 30% which may be interpreted as the proportion of amylose complexed with lipid and hence unavailable to I_2 (assuming minimal displacement of lipid by I_2). A similar value may be calculated on the basis of the following assumptions: amylose present as a helix of 6 glucose residues per unit and period of 8 Å between repeat units;[27] C_{18} linear acyl chain (approximately 20 Å) is aligned in helix; lipid approximately 1% of starch (4% by weight of amylose).

The acyl chain of the amylose-lipid complex is usually represented as a straight "rod-like" structure. However, approximately 50% of the fatty acid chains are of linoleic acid (cis-9,cis-12-18:2). It might be expected that the 'kinks' introduced by two cis-double bonds would interfere with complex formation. Although there is some evidence that the complex between amylose and linoleic acid is slightly less stable than that between amylose and the corresponding fully saturated fatty-acid, the free energies determining the conformation of the complexes with saturated or unsaturated fatty acids are, presumably, relatively similar.

It was shown earlier (Fig. 1) that starches from different sources demonstrate very different viscosity behaviors when heated in water. The effects of internal (and surface) lipids on the physical behavior of starches have not yet been fully evaluated. However, preliminary experiments in our laboratories[22] have clearly shown that removal of lipid from cereal starches is accompanied by significant changes in viscosity characteristics of the starch-water pastes and that addition of lipids from maize or wheat starch to potato starch (which has no endogenous lipid) markedly affects the viscosity properties of the potato starch paste. In attempting to investigate the role of internal components (lipid and protein) of starch granules on physical properties, enzyme susceptibility, etc., one is faced by the major problem that it is not possible to remove these components without disrupting the organization of the starch granule.

During the gelatinization process, starch granules first take up water, swell and eventually release the granule contents into gel or solution. The rate and extent of swelling varies with the source of starch and both surface and internal components play a role in determining the interaction with water. In attempts to explain differences between swelling behavior of different starches my colleagues[23,24] have carefully examined starch preparations by a range of microscope techniques. Large, lenticular starch granules from the Triticeae (wheat, barley, rye) all undergo a very similar swelling process (Fig. 6) that appears to involve only two-dimensional swelling in the plane of the lenticular granules and then a constraint on further radial swelling and little increase in thickness of the granule; subsequent expansion appears to be by a puckering at the periphery resulting in retention of the 2-dimensional effect and a relatively constant thickness of the granule. In this respect, we are unable to distinguish between wheat, barley and rye (which have similar analytical features with respect to amylose, amylopectin, lipid) but no other starches examined to date appear to show this effect; all others examined show a general three-dimensional swelling effect.

Distribution of Minor Components Within Starch Granules

There is come controversy over the distribution of amylose and amylopectin within starch granules[25] which is reflected in the uncertainty over the mode of starch deposition in developing granules. An increase in the proportion of amylose during starch synthesis in developing endosperm is generally observed[1,2] and thus a greater concentration of amylose near the periphery of large granules might be expected. A similar gradient in lipid concentration should give higher lipid to starch ratios in granules from germinating grains in which the outer region of starch granules appear to be less readily hydrolyzed. Such analyses with wheat starch showed a slightly increased concentration of lipid in the outer regions of partially-degraded granules prepared from germinating wheat as compared with intact

Fig. 6. Scanning electron micrographs of lenticular starch granules from wheat grain. Suspensions of starch were heated for 10 min at the following temperatures: (a) 20°C; (b) 40°C; (c) 50°C; (d) 60°C; (e) 70°C; (f) 80°C; (g) 90°C; (h) 97°C. Data from Bowler et al,[23] reproduced with permission.

granules from ungerminated wheat grains.[12] In a preliminary examination of wheat starch granules by X-ray fluorescence electron microscopy, Morrison and Milligan[8] report evidence for a gradient of phosphorus (presumed to be present predominantly as lysophosphatidylcholine) distribution, increasing from the center to the periphery of the granules.

Lipid analyses on wheat starch granules of different sizes from two varieties of wheat led Meredith et al[26] to conclude that the amounts of lipid per granule were proportional to the surface area of large granules, but proportional to the volume of small granules; this may indicate differences in the deposition mechanisms in the two types of granules and further clarification is indicated.

STARCH LIPIDS IN DEVELOPING GRAIN

The relationships between starch lipids and amylose have been noted above. It is also known that the lipid content of starch granules increases during starch deposition in developing grains of barley[28] and maize;[29] an increase in the proportion of lysophospholipid in the total lipids also occurs during the development of wheat grains.[30] The close relationship between starch accumulation and starch lipid content suggests a direct role for the lipids in starch synthesis and granule development.[8] Other groups have reached similar conclusions; for example, Vieweg and de Fekete[31] have proposed major roles for starch lipid in the metabolism (both the biosynthesis and degradation) of starch; Downton and Hawker[32] claimed that lysophospholipids are important in controlling starch synthesis through the starch synthetase enzyme system on the grounds that a lipid requirement for enzyme activity was observed and, in chilling-sensitive plants, a biphasic Arrhenius plot of starch synthesis vs 1/T was obtained, suggesting a role in the starch synthesis for membrane-bound unsaturated lipids.

However, the direct involvement of internal starch lipids (monoacyl lipids) in starch synthesis must be questionable. If it is assumed that biosynthesis of starch (amylose and amylopectin molecules and the relative proportions of these) has common features in plants, then the absence or low concentration of starch internal lipids in many starches (see Table 4) must be explained; the starch internal monoacyl lipids appear to be peculiar to cereal starch granules. Moreover, the nature of the internal lipids differs between cereal species (see Tables 2 and 5). Wheat starch contains almost exclusively lysophospholipids whereas in maize and rice starches free fatty acids and lysophospholipids are important lipids (Table 5). Thus any requirement for lipid in amylose synthesis would appear to be non-specific. The cytotoxic and membrane disrupting nature of the typical starch lipids (lysophospholipids and free fatty acids) also argues against any role in metabolic processes in which these lipids have access to membrane systems.

Starch biosynthesis within the amyloplasts of developing tissue usually involves membrane-bound enzyme systems.[33] It is intriguing that in waxy maize mutants (low amylose, low starch internal lipid; see Table 6) the starch synthetase enzyme system is soluble and the particulate form (typical of the normal, amylose-containing varieties) is absent; moreover the soluble enzyme system from waxy maize will bind to normal, amylose-containing starch granules and to isolated amylose. Available evidence points to a possible role for diacyl lipids, as components of lipo-protein membrane systems in starch biosynthesis; these membrane diacyl lipids should, however, be distinguished from the membrane-disrupting monoacyl lipids present within starch granules. However, it seems more probable that monoacyl lipids associated with amylose in cereal starches are hydrolysis products from lipids of lipo-protein membranes involved in starch biosynthesis and are rendered metabolically inactive by becoming bound to amylose chains. In tissues where starch granules do not contain significant quantities of internal lipid, it is possible that the amyloplast and its internal membrane systems are degraded more rapidly and extensively during starch deposition such that the lipids are not present in the mature granules. Certainly, potato tubers are known to contain high levels of lipolytic and lipid oxidizing enzymes that catalyze the breakdown of mono- and di-acyl lipids of membrane systems.[34] The corresponding lipid-degrading enzymes in cereals, although present, appear to be less active.[35]

The above discussion refers to postulates made concerning biosynthetic mechanisms but based upon <u>analytical</u> data; <u>metabolic</u> studies on starch lipid complexes in developing grains have been few. Lysophosphatidylcholine, the major starch lipid of wheat and barley could be synthesized by acylation of a glycerol ester precursor; or formed by partial deacylation or transacylation from (diacyl) phosphatidylcholine (Fig. 7). Attempts to identify the mechanisms of starch lipid formation have been made in the reviewer's laboratory and in collaboration with others. Baisted[36] examined the incorporation of ^{14}C-choline into phosphatidylcholine (PC) and lysophosphatidylcholine (lysoPC) in endosperm tissue from developing grains of barley. He showed that labeling of lysoPC reached a maximum when the seeds had attained 60 to 70% of their maximal weights and that starch-bound lysoPC in isolated ears of grain continued to accumulate label after 72 h, whereas 'free' lysoPC activity declined after about 50 h. The results indicated that different 'pools' of lysoPC were present. Stokes et al,[30] working with developing wheat grains, showed an increase in the proportion of lysoPC in wheat endosperm during development. They were unable to detect incorporation of label from ^{14}C-oleic acid into the starch lipid at early stages of development (5-10 days after anthesis) although the label was incorporated into other, non-starch lipids. Further work is required to establish the mechanisms of starch lipid formation and complexing.

LIPID COMPLEXES AND STARCH DEGRADATION

The reactions of amylose-lipid complexes have been studied quite extensively because the complexes are relatively easy to prepare on a large scale and they can be designed to have specific properties, both functional and nutritional. One property to receive attention has been the resistance of

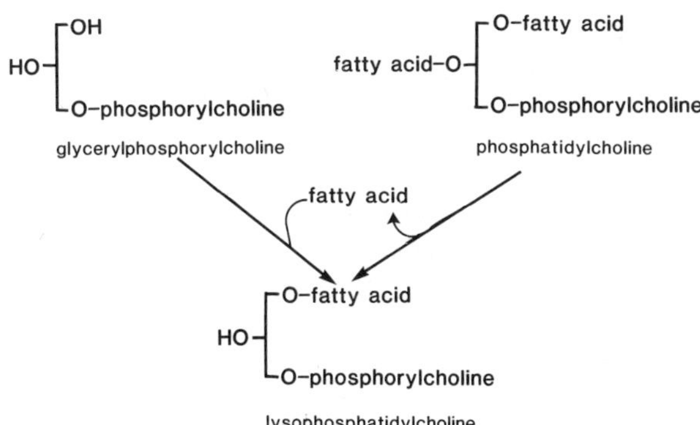

Fig. 7. Simplified scheme to show alternative routes for the biosynthesis of lysophosphatidylcholine.

amylose to enzymic degradation when complexed with lipid. Mercier et al[37] have proposed that starch may be rendered less digestible, reducing its calorific value and increasing its 'dietary fiber' contribution if its amylose content is complexed and made unavailable to the amylolytic enzymes in the upper digestive tract. Larsson and Miezis[38] suggested that a significant amount of dietary amylose, initially in an uncomplexed form, may form lipid complexes with the monoacyl lipids normally produced by pancreatic lipase action on dietary fats.

In view of the possibility that 'synthetic' amylose-lipid complexes may be resistant to amylolytic attack, it is of interest to ascertain the fate of amylose associated with lipid in starch granules of germinating seeds. Some recent studies by my colleague Jim Sargeant (Table 7) have shown that the amylose-amylopectin ratios are similar in undegraded and partly (60%) degraded wheat starch; this indicates that the amylose and amylopectin components are degraded at similar rates, although results of other studies with rice starch[39] have been claimed to show preferential hydrolysis of amylopectin. It would also seem that there is not a great difference between the rates at which uncomplexed amylose and lipid-associated amylose in cereal starches are degraded during germination. Table 7 shows that the lipid contents of intact and partly degraded starch are quite similar and the slightly increased proportion of lipid in the starch from partly germinated wheat could be anticipated

Table 7. Analysis of large, lenticular starch granules from ungerminated and germinated wheat grain.

	Amylose:amylopectin ratio	Starch internal lipid content
	%	mg·100 g^{-1} starch
Ungerminated grain	28.4:71.6	451
Germinated 5 days, 20°C	29.6:70.4	511

Data from J. G. Sargeant.[12]

if in fact there is a higher concentration of lipid in the outer layers of large lenticular starch granules from wheat (see above). Recent studies in Baisted's laboratory (personal communication) have also shown that for both a normal and a high amylose variety of barley, the lysophospholipid:amylose ratios do not change during germination, indicating that the starch-degrading enzymes do not distinguish between amylose and its lipid complex.

Baisted's group [40,41] has also studied the relationships between starch and phospholipids of barley and their enzymic degradation during germination. Some of their results, summarized in Fig. 8, show a close time relationship between the loss of starch-bound lysoPC from the endosperm and the activity of an acyl hydrolase enzyme that attacks lysoPC and that is present in the endosperm in both particulate and soluble forms. The increase in endosperm α-amylase activity

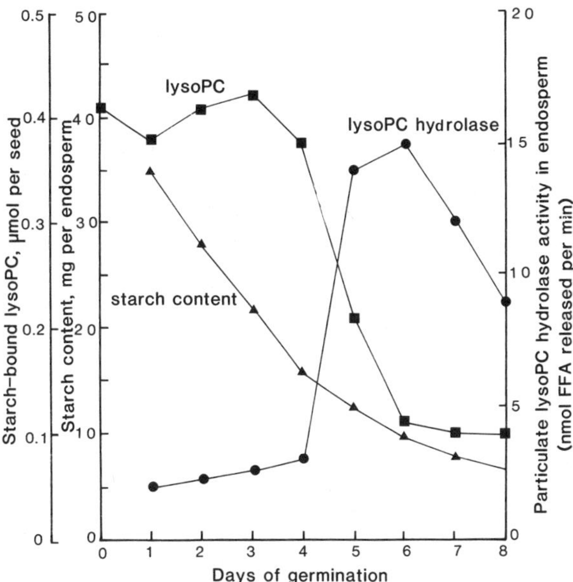

Fig. 8. Levels of starch, starch-bound lysophosphatidyl-choline (lysoPC) and lysoPC-acyl hydrolase in endosperm of germinating barley (adapted from Baisted[40] and Baisted and Stroud[41]).

(not shown) parallels that of the lipid acyl hydrolase. There is no accumulation of free lysoPC as the starch is degraded; presumably the acyl hydrolase acts to remove the potentially damaging lysophospholipid as it is released from the amylose complex. Further work from this group[42] has recently demonstrated the effects of gibberellic acid in stimulating lipid acyl hydrolase activity in barley.

It is clear from the discussion above that, although the close association of amylose and monoacyl lipids in cereal starches is well established, the role (if any) of these lipids in the degradation of starch is by no means clear.

There seem to be significant differences in the ways in which starch-degrading enzymes attack amylose which is artifically complexed with lipid compared with amylose that is naturally associated with lipid in cereal starch granules. Further research on the effects of starch-lipid interactions in the degradation of starch granules is obviously necessary.

CONCLUSIONS

In considering the functions and metabolism of 'starch', the chemical and biological definitions should be distinguished, and the possible roles of minor components of starch granules considered. Proteins, lipids and other components are associated with starch granules isolated from cereal grains. Some of these are present on or near the surface of the granules; others are buried within the granule matrix and are only extracted under conditions that cause disruption of granule structure.

Wheat starch, isolated in aqueous media, contains an unusual protein species, assumed to be at the surface of the granule in addition to other proteins extractable only under dissociating conditions. The location and functions of these proteins in wheat starch granules _in situ_ are not known.

Isolated starch granules from cereal grains contain lipids that are extractable in cold polar solvents and which are assumed to be associated with the granule surface or with endosperm components contaminating the starch granule preparations. The composition of this lipid indicates that it may include amyloplast membrane degradation products.

The presence of lipids that are less readily extractable and that are assumed to be within the granule matrix, is now well established for those cereal grains examined to date. For cereal starches the internal lipid amounts to about 1% of the total starch dry weight. It is significant that, of all the starches carefully analyzed, only those starches from cereal grains contain significant amounts of internal lipid; tuber, root and legume seed starches contain much lower lipid levels than cereal grain starches. The lipids within cereal starch granules are monoacyl lipids (lysophospholipids and free fatty acids are the main internal lipid components of cereal starches examined to date). Such lipids, which are cytotoxic and not usually found in healthy plant tissues, are able to form complexes with amylose. Amylose-lipid complexes contain a single fatty acyl chain within the core of an α-glucan helix. Although such complexes can be formed readily by appropriate treatment of starch, evidence that such complexes occur naturally in cereal starches is not yet fully established.

Recently, roles for starch lipids have been suggested in the biosynthesis and catabolism of starch, particularly in relation to the differences in the metabolism of the mainly linear amylose fraction and the branched amylopectin. However, direct evidence for such roles for starch lipids *in vivo* is wanting and the apparent restriction of internal lipids (at least in mature granules) to certain plant species only, lowers the probability of a general metabolic role for starch lipids.

ACKNOWLEDGMENTS

I am grateful to Professors W. R. Morrison of the University of Strathclyde and D. J. Baisted of Oregon State University for supplying me with material prior to publication. Research support from the Director of Research, R.H.M. Research Ltd. and from Ministry of Agriculture, Fisheries and Food is gratefully acknowledged.

REFERENCES

1. BANKS W, CT GREENWOOD 1975 Starch and Its Components. Endinburgh University Press, Edinburgh
2. BANKS W, DD MUIR 1980 Structure and chemistry of the starch granule. *In* J Preiss, ed, The Biochemistry of

Plants, Vol 3, Carbohydrates: Structure and Function, Academic Press, New York pp 321-369

3. STENVERT NL, K KINGSWOOD 1977 Influence of physical structure of protein matrix on wheat hardness. J Sci Fd Ag 29: 11-19
4. KASSENBECK P 1978 Interactions of starch and protein during dough formation. 6th World Cereal and Bread Congress, Winnipeg, Abst, Cereal Foods World 23: 494
5. SIMMONDS DH, KK BARLOW, CW WRIGHT 1973 The biochemical basis of grain hardness in wheat. Cereal Chem 50: 553-563
6. MORRISON WR 1981 Starch lipids: a reappraisal. Starch 33: 408-410
7. FISHWICK MJ, AJ WRIGHT 1980 Isolation and characterization of amyloplast envelope membranes from _Solanum tuberosum_. Phytochemistry 19: 55-59
8. MORRISON WR, TP MILLIGAN 1982 Lipids in maize starches. _In_ GE Inglett, ed, Maize: Recent Progress in Chemistry and Technology, Academic Press, New York pp 1-18
9. HARGIN KD, WR MORRISON 1980 The distribution of acyl lipids in the germ, aleurone, starch and non-starch endosperm of four wheat varieties. J Sci Fd Ag 31: 877-888
10. HARGIN KD, WR MORRISON, RG FULCHER 1980 Triglyceride deposits in the starchy endosperm of wheat. Cereal Chem 57: 320-325
11. LOWY GDA, JG SARGEANT, JD SCHOFIELD 1981 Wheat starch granule protein: the isolation and characterization of a salt-extractable protein from starch granules. J Sci Fd Agric 32: 371-377
12. SARGEANT JG 1980 α-Amylase isoenzymes and starch degradation. Cereal Res Commun 8: 77-86
13. MARSH RA, SG WAIGHT 1982 The effect of pH on the zeta potential of wheat and potato starch. Starch 34: 149-152
14. BADENHUIZEN NP 1969 The Biogenisis of Starch Granules in Higher Plants. Appleton-Century-Crofts, New York
15. MIKUS FF, RM HIXON, RE RUNDLE 1946 The complexes of fatty acids with amylose. J Amer Chem Soc 68: 115-1123
16. ACKER L 1977 The lipids of starch: research between carbohydrates and lipids. Fette Seifen Anstrichmittel 79: 1-9
17. MORRISON WR, B LAIGNELET 1983 An improved method for determining apparent and total amylose in cereal and other starches. J Cereal Sci 1: 9-20

18. SARGEANT JG 1982 Determination of amylose: amylopectin ratios of starches. Starch 34: 89-92
19. ITO S, S SATO Y FUJINO 1979 Internal lipid in rice starch. Starch 31: 217-221
20. TAN SL, WR MORRISON 1979 The distribution of lipids in the germ, endosperm, pericarp and tip cap of amylomaize, LG-11 hybrid maize and waxy maize. J Amer Oil Chem Soc 56: 531-535
21. MORRISON WR 1978 The stability of wheat starch lipids in untreated- and chlorine-treated cake flours. J Sci Fd Agric 29: 365-371
22. MELVIN MA 1979 The effect of extractable lipid on the viscosity characteristics of corn and wheat starches. J Sci Fd Ag 30: 731-738
23. BOWLER P, MR WILLIAMS, RE ANGOLD 1980 A hypothesis for the morphological changes which occur on heating lenticular wheat starch in water. Starch 32: 186-189
24. WILLIAMS MR, P BOWLER 1982 A morphological study of triticaceous and other starches. Starch 34: 221-223
25. MEREDITH P 1981 Large and small starch granules in wheat - are they really different? Starch 33: 40-44
26. MEREDITH P, WR MORRISON HN DENGATE 1978 The lipids of various sizes of wheat starch granules. Starch 30: 119-125
27. CARLSON TL-G, K LARSSON, N DINH-NGUYEN, N KROG 1979 A study of the amylose-monoglyceride complex by Raman spectroscopy. Starch 31: 222-224
28. BECKER G, L ACKER 1972 Lipids of barley starch and their alteration during the growth of barley. Fette Seifen Anstrichmittel 74: 324-328
29. TAN SL, WR MORRISON 1979 Lipids in the germ, endosperm and pericarp of the developing maize kernel. J Amer Oil Chem Soc 56: 759-764
30. STOKES DN, T GALLIARD, JL HARWOOD 1980 Lipid metabolism in developing wheat seed. In P Mazliak, P Benveniste, C Costes, R Douce, eds, Biogenesis and Function of Plant Lipids, Elsevier-North Holland, Amsterdam pp 223-226
31. VIEWEG GH, MAR de FEKETE 1980 On the effect of lipids on starch-metabolizing enzymes and its significance in relation to the simultaneous synthesis of amylose and amylopectin in starch granules. In JJ Marshall, ed, Mechanisms of Saccharide Polymerization and Depolymerization, Academic Press, New York pp 175-185

32. DOWNTON WJS, JS HAWKER 1975 Evidence for lipid-enzyme interaction in starch synthesis in chilling-sensitive plants. Phytochemistry 14: 1259-1263
33. PREISS J, C LEVI 1980 Starch biosynthesis and degradation. In J Preiss, ed, The Biochemistry of Plants, Vol 3, Carbohydrates: Structure and Function, Academic Press, New York pp 371-423
34. GALLIARD T 1980 Degradation of acyl lipids: hydrolytic and oxidative enzymes. In PK Stumpf, ed, The Biochemistry of Plants, Vol 4, Lipids: Structure and Function, Academic Press, New York pp 85-116
35. GALLIARD T, PJ BARNES 1980 The biochemistry of lipids in cereal crops. In P Mazliak, P Benveniste, C Costes, R Douce, eds, Biogenesis and Function of Plant Lipids, Elsevier-North Holland, Amsterdam pp 191-198
36. BAISTED DJ 1979 Lysophosphatidylcholine biosynthesis in developing barley. Phytochemistry 18: 1293-1296
37. MERCIER C, R CHARBONNIERE, D GALLANT, A GUILBOT 1979 Structural modification of various starches by extrusion-cooking with a twin-screw French extruder. In JMV Blanshard and JR Mitchell, eds, Polysaccharides in Food, Butterworths, London pp 153-170
38. LARSSON K, Y MIEZIS 1979 On the possibility of dietary fibre formation by interactions in the intestine between starch and lipids. Starch 31: 301-302
39. FUKUI T, Z NIKUNI 1956 Degradation of starch in the endosperm of rice seeds during germination. J Biochem (Tokyo) 43 (1): 33-40
40. BAISTED DJ 1981 Turnover of starch-bound lysophosphatidylcholine in germinating barley. Phytochemistry 20: 985-988
41. BAISTED DJ, F STROUD 1982 Soluble and particulate lysophospholipase in the aleurone and endosperm of germinating barley. Phytochemistry 21: 29-31
42. BAISTED DJ, F STROUD 1982 Enhancement by gibberellic acid and asymetric distribution of lysophospholipase in germinating barley. Phytochemistry 21: 2619-2623

Chapter Seven

THE COOPERATIVE ROLE OF ENDO-β-MANNANASE, β-MANNOSIDASE AND α-GALACTOSIDASE IN THE MOBILIZATION OF ENDOSPERM CELL WALL HEMICELLULOSES OF GERMINATED LETTUCE SEED

J. DEREK BEWLEY, DAVID W. M. LEUNG AND FRANCIS B. OUELLETTE

Plant Physiology Research Group
Department of Biology
University of Calgary
Calgary, Alberta T2N 1N4 Canada

INTRODUCTION

The embryo of the lettuce 'seed' (which, strictly, is an achene), is completely surrounded by an endosperm comprised of only two or three cell layers.[1,2] These cells have thick walls,[2] and the endosperm acts as a barrier to germination, i.e. it imposes dormancy, when the intact seed is incubated in darkness at supra-optimal temperatures (ca. 25°C). A number of factors will promote the seed to overcome this dormancy, and particularly well-studied have been the red light stimulus mediated via the phytochrome system,[3,4] and the promotive effect of gibberellin when applied in darkness.[5,6]

Following penetration of the germinated radicle through the endosperm, degradation of the cell walls of this latter structure commences.[2] Complete hydrolysis of the endosperm can be achieved within 15 to 20 h of radicle emergence,[7] and it occurs prior to the mobilization of the major reserves (lipid, protein and phytate) stored within the cotyledons.[7-9] Hence, early axis growth appears to be independent of a supply of hydrolytic products from the cotyledons but, instead, may utilize the carbohydrate source provided by degradation of the endosperm cell walls. Our interest has been to determine the nature of the carbohydrates made available by the breakdown of the endosperm cell wall, the enzymes involved in this hydrolytic process, and the endogenous control mechanisms which operate to ensure that endosperm hydrolysis occurs exclusively as a post-germination phenomenon, but prior to mobilization of the major reserves. This chapter reviews our progress so far.

HYDROLYSIS OF THE ENDOSPERM CELL WALL

Cell Wall Composition

Analysis of the composition of the cell walls of the radicle and endosperm of lettuce seed reveals that whereas the former is largely cellulosic in nature (i.e. the predominant sugar is glucose), the latter contains a high proportion of mannose[10] (Table 1). Thus the endosperm cell wall is composed to a large extent of mannose-containing polysaccharides--probably (1→4)-β-mannans. Some galactose is also found within these cell walls, which is suggestive of the presence of galactomannans, although the nature of the carbohydrate linkages within the lettuce endosperm cell wall remains to be elucidated.

Table 1. Percentage of sugar composition of lettuce seed tissue cell walls (mg per 100 mg carbohydrate).[a]

Sugar	Endosperm	Radicle
Rhamnose	0.4	1.7
Arabinose	9.5	1.5
Xylose	1.5	6
Mannose	58	7.4
Galactose	9.5	8.6
Glucose	10	28
Uronic acids	10	28

[a] Based on data in Halmer et al[10]

Hydrolysis of polymeric mannans to mannose requires the presence of two enzymes, endo-β-mannanase, and β-mannosidase:

$$\text{Mannans} \xrightarrow{\text{endo-}\beta\text{-mannanase}} \text{Oligomannans}$$

$$\text{Oligomannans} \xrightarrow{\beta\text{-mannosidase}} \text{Mannose}$$

If galactose is present, then this must be cleaved by the enzyme α-galactosidase.

endo-β-Mannanase

Extensive studies by Peter Halmer and co-workers[8,10-12] have revealed that germinated lettuce seeds contain up to 100-times more endo-β-mannanase than freshly imbibed ones. This production occurs only in germinated seeds, regardless of the treatment applied to promote their germination. Since the work on mannanase has already been reviewed,[7] we will confine ourselves here to summarizing the major points of interest. The enzyme is produced exclusively within the endosperm, as a post-germinative event, although its synthesis requires the participation of both the radicle and the cotyledon portions of the embryo. Their involvement is summarized in Fig. 1. Essentially, within the first few hours following irradiation with red light, and long prior to the completion of germination, some readily diffusible

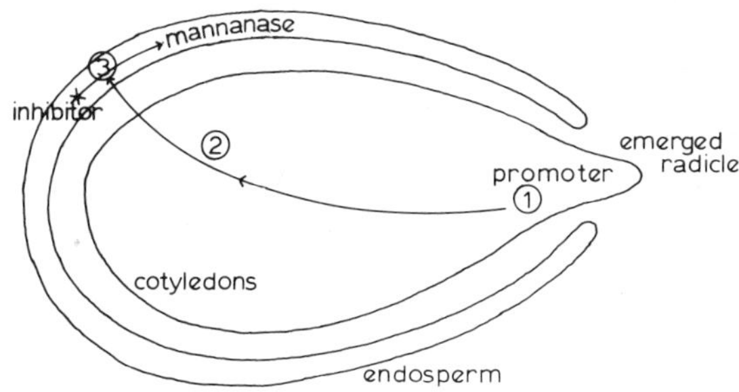

Fig. 1. A hypothetical scheme to demonstrate the interrelationships between radicle, cotyledons and endosperm in the synthesis of endo-β-mannanase in red light-induced lettuce seeds cv. Grand Rapids. See text for details.

promotive substance(s) is produced within and/or released from the radicle into the cotyledons (event 1, Fig. 1). The nature of this promoter is not known, but experimentally it can be replaced by low levels of gibberellin and cytokinin, which is suggestive (but by no means proof) of their natural involvement. Within the cotyledons the promoter is modified, or sets some other event in motion to produce a different promoter (event 2, Fig. 1), prior to its transfer to the endosperm. That the cotyledons must be involved was shown in experiments where either the unknown promoter from the axis, or applied gibberellin and cytokinin failed to stimulate mannanase production within isolated endosperms, but would do so only if all, or part, of the cotyledons was present.[11] The transfer of the promotive substance(s) to the endosperm occurs at about the time of radicle emergence. Hence delay in the passage of (modified) promoter from the radicle to the endosperm until germination is completed may be the effective control mechanism which ensures that mannanase production is a post-germinative event.

Synthesis of mannanase within the endosperm is possible at early stages following imbibition, many hours before germination is completed, but this is prevented by the presence of inhibitor(s) within this tissue. As with the promoter, the nature of the inhibitor is as yet undetermined, although its activity can be mimicked by exogenously applied abscisic acid. The promotive substance diffusing from the cotyledon overcomes the effect of the inhibitor (event 3, Fig. 1) and allows mannanase synthesis to proceed.

Thus, in brief summary, endo-β-mannanase synthesis within the endosperm is under the control of the embryo. Its synthesis is promoted by red light and certain chemical (hormonal) stimulants of germination. The control mechanism of enzyme synthesis involves the overcoming of the effect of an inhibitor present in the endosperm by promoter from the cotyledons, produced in response to events initiated by the axis.

β-Mannosidase

The activity of this enzyme contrasts markedly with that of endo-β-mannanase in three respects: (a) it is present exclusively within the cotyledons, (b) it is not a readily soluble or diffusible enzyme, and (c) it is present in appreciable quantities in both dry and imbibed, dormant seeds (Fig. 2). Extraction of cotyledons with 20 mM phosphate buffer (pH 7) yields a fraction of 'salt-soluble', or readily soluble activity, which is negligible in the dry seed and remains inappreciable in the imbibed seed. Considerably more activity is obtained when the fraction that is insoluble in this low molarity buffer is re-extracted with buffered 2 M NaCl. This activity, combined with the still insoluble activity of the resultant pellet has been termed the 'bound' activity, and is probably indicative of β-mannosidase closely associated with, or bound to, the cell wall fraction of the cotyledons. It is evident from Fig. 2 that neither the low salt-soluble nor the bound activities increase during or following germination of light-stimulated seeds, compared either to the dark controls, or to the unimbibed seed. Obviously, then, β-mannosidase is present within the lettuce seed when there is no substrate available for it to utilize. In red light-stimulated seeds, endo-β-mannanase releases oligomannans as hydrolytic products from

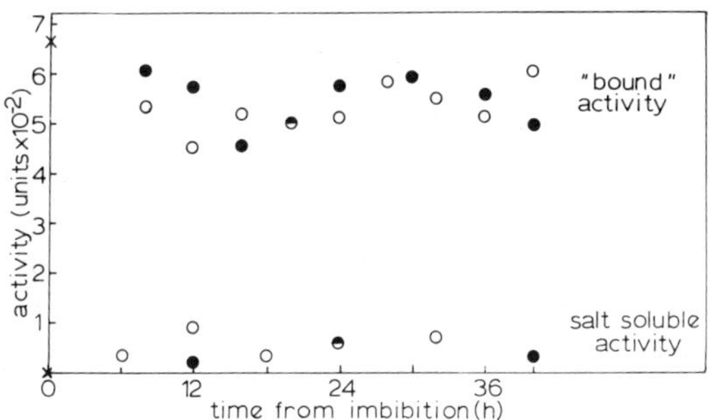

Fig. 2. The activity of high- ("bound" activity) and low (salt-soluble activity) salt extractable β-mannosidase from dry seeds (**X**), seeds incubated in darkness (●) and seeds irradiated with 5 min red light after a 2-h-imbibition period (O). Salt soluble activity was extracted with 20 mM phosphate buffer (pH 7), and the resultant pellet extracted for bound activity by the addition of 2 M NaCl in buffer. The supernatant and pellet after high salt extraction were assayed separately and their activities summed to provide the value for bound activity. The assay mixture included p-nitrophenyl-β-D-mannopyranoside[18] (Sigma) in McIlvaine buffer (pH 4.8), and 1 unit of activity is defined as the number of pmoles of p-nitrophenyl released per min per cotyledon pair.

the endosperm cell wall. It seems reasonable to expect that these then diffuse across the fluid-filled space between the endosperm and the cotyledons, to be subsequently hydrolyzed by β-mannosidase residing within the cell walls of the latter. The resultant product, mannose, is presumably then readily absorbed into the peripheral cells of the cotyledons and therein utilized.

It should be noted here that in a previous publication[9] we reported an increase in what is the equivalent of 'salt soluble' activity in cotyledons of red light-stimulated seeds. We now believe that observation to be incorrect, although the discrepancy between the results presented here, and those reported previously is difficult to explain. Of several possibilities, two appear to be plausible: (a) the conditions of assay previously used did not rule out that an increase in contaminating substances (e.g. phenolics) within the enzyme extract produced or released following red light treatment could have interfered with the sensitive assay system used, or (b) the chromogenic substrate, *p*-nitrophenyl-β-*D*-mannopyranoside, used in previous assays was laboratory manufactured and could have contained impurities which reacted to other enzymes or compounds present in the low salt buffer extract.

α-Galactosidase

This enzyme, like β-mannosidase, is present in the dry seed, but it increases in activity following stimulation by red light.[13,14] In contrast to mannanase, however, it increases during germination, and many hours prior to emergence of the radicle through the endosperm (Fig. 3). The largest increase in activity occurs within the cotyledons, although some does occur within the endosperm.[14] Since this enzyme is readily diffusible, however, it is not clear at the present time whether the elevation in activity in the endosperm is an *in situ* event, or whether it results simply by diffusion of increased enzyme from the cotyledons.

The action of phytochrome in inducing α-galactosidase production is extremely rapid, it being completed within a matter of seconds.[13] On the other hand, increased enzyme activity does not occur until some 2 h after irradiation (Fig. 3). This is because the site of perception of the red light is within the axis, and the site of increased enzyme activity is the cotyledons (and perhaps the surrounding

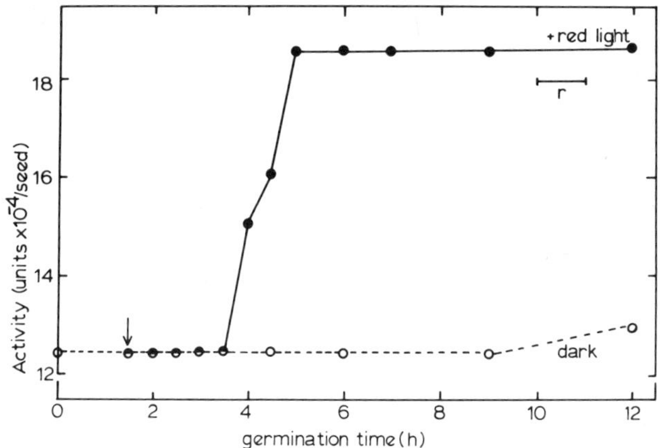

Fig. 3. α-Galactosidase activity in extracts of dark-imbibed (○) and red-light irradiated (●) lettuce seeds during germination. The assay procedure is as Leung and Bewley.[14] One unit of activity is defined as the number of nmoles p-nitrophenyl released per min per seed. The first seeds completed germination 10 to 11 h from the start of imbibition (denoted by r).

endosperm). The delay is probably because of the time required for the light-induced stimulus, which is a translocatable substance of unknown nature, to pass from the axis to the cotyledons.[14] Interestingly, as in the case of mannanase induction, both gibberellin and cytokinin can substitute for the axis.[14]

Isolated endosperms, i.e. those dissected from imbibed seeds and incubated in volumes of water of at least 400 μl per 20 endosperms (a 'large' volume) undergo extensive cell wall degradation, as shown by the release of sugars into the incubation medium, and produce endo-β-mannanase (Fig. 4). Of the sugars that are released, one that increases substantially in the medium is galactose (Fig. 4). This suggests the involvement of α-galactosidase in cell wall hydrolysis. That indeed it is involved is shown in the following sequence of experiments.

myo-Inositol is a competitive inhibitor of α-galactosidase[15] (Fig. 5) and it drastically reduces the in vivo activity of this enzyme in isolated endosperms incubated in a large volume (Table 2). Although a concentration of 250 mM *myo*-inositol has to be used to obtain maximum inhibiton, there is no resultant osmotic effect on the endosperm cells because mannitol at an equivalent concentration is uninhibitory to enzyme production. The reduction in α-galactosidase activity due to *myo*-inositol is mirrored by a reduction in galactose release from the isolated endosperms (Fig. 4), the inhibition being 63%. That α-galactosidase is involved in the hydrolysis of endosperm cell walls can also be shown by comparing the hydrolytic products released in the presence and absence of α-galactosidase activity. Isolated endosperms incubated in a large volume in the absence of *myo*-inositol, i.e. allowing both endo-β-mannanase activity (which is unaffected by the inhibitor) and α-galactosidase activity, yield oligomers separable by Bio-Gel chromatography into sizes approximately equivalent to single glucose units (G_1) and tetramers (G_4), with lesser amounts of oligomers of larger size being produced also (Fig. 6, minus *myo*-inositol). Similarly incubated isolated endosperms, but with *myo*-inositol present, yield less of the monomer and small oligomer units (G_1 and G_4), but release from the cell walls a predominant large oligomer, equivalent in size to at least 20 to 25 glucose units (Fig. 6, plus *myo*-inositol). We can conclude that in the absence of α-galactosidase activity the final processing of the large oligomers (G_{20-25}) released from the wall by mannanase activity does not take place. It is possible, then, that both enzymes are required to produce

Fig. 4. Production of endo-β-mannanase (**x**), sugar released from endosperm cell walls (●) and release of galactose into the incubation medium (○) in endosperms isolated from the intact seed after 4 h from the start of imbibition in darkness, and incubated in a 'large' volume. The histogram shows the amount of galactose released when endosperms were incubated in 250 mM *myo*-inositol. The assays for endo-β-mannanase and sugar release from the cell wall are detailed in Halmer et al[10] and Halmer and Bewley.[11] Released galactose was determined by enzyme microassay using galactose dehydrogenase.[19]

Fig. 5. Lineweaver-Burke plot of the effect of *myo*-inositol on activity of extractable lettuce seed α-galactosidase, to illustrate a classical competitive inhibition. α-Galactosidase assay as in Leung and Bewley.[14]

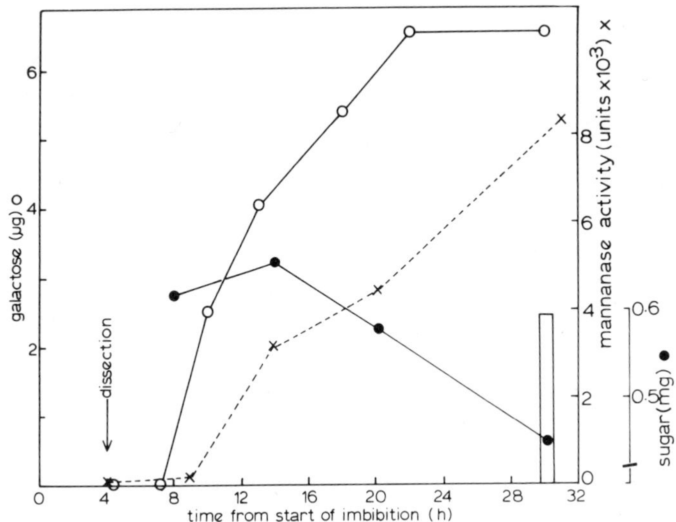

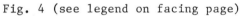

Fig. 4 (see legend on facing page)

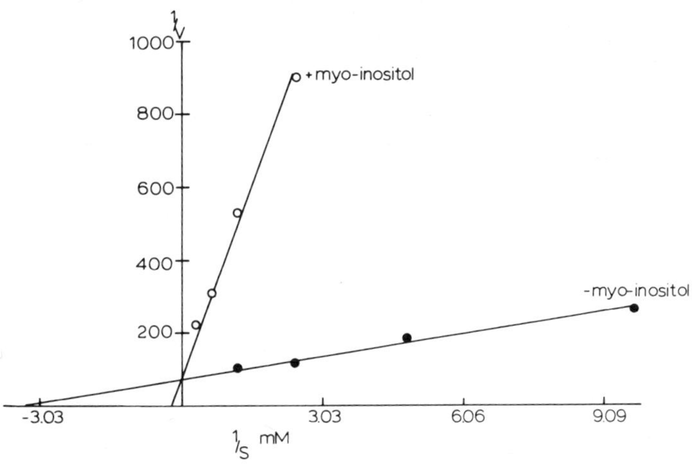

Fig. 5 (see legend on facing page)

Table 2. Effect of myo-inositol on *in vivo* α-galactosidase activity of isolated lettuce endosperms during incubation in a large volume.

	p-Nitrophenol formed per 100 μl 'endosperm diffusate'	Inhibition
	nmol	%
- myo-Inositol	14.2	0
+ myo-Inositol, 250 mM	4.2	70
+ Mannitol, 250 mM	15.0	0

Incubation time: 26 h from dissection in darkness at 23°C. Assay procedure as in Leung and Bewley.[14]

oligomers of sufficiently small size for β-mannosidase to act upon.

The nature of the linkages cleaved by α-galactosidase to yield the small oligosaccharides is not known. But if the galactomannans are of the type found in the endosperms of leguminous seeds, the galactose will be linked by (1→6)-α bonds as unit side chains to a long backbone of (1→4)-β-linked mannose residues.[16] As outlined below, the mannanase may first be required to hydrolyze the native galactomannans to generally large (20-25 residue) mannose oligomers containing galactose side-chains. Before further hydrolysis can proceed, the galactose units may have to be removed by α-galactosidase, whereafter final hydrolysis can be completed by the mannanase.

That α-galactosidase is incapable of releasing galactose from the native cell wall, but only from the products of prior endo-β-mannanase activity is shown in Table 3. Previous studies[7,11] have shown that incubation of isolated endosperms in a 'small' volume (200 μl or less per 20 endosperms) prevents mannanase production, due to the presence of endogenous inhibitor (see Cell Wall Composition section). This is reiterated in Table 3, where mannanase activity produced by endosperms incubated in a small volume is less than 5% of that produced in a large volume. Reduction of the incubation volume has no effect upon α-galactosidase activity, on the other hand. The release of galactose from the endosperms into the surrounding medium occurs in the large volume, and not the small, even though α-galactosidase activity is present in the latter (Table 3). Thus, in the absence of mannanase activity, no galactose is released, showing that this sugar can only be produced when α-galac-

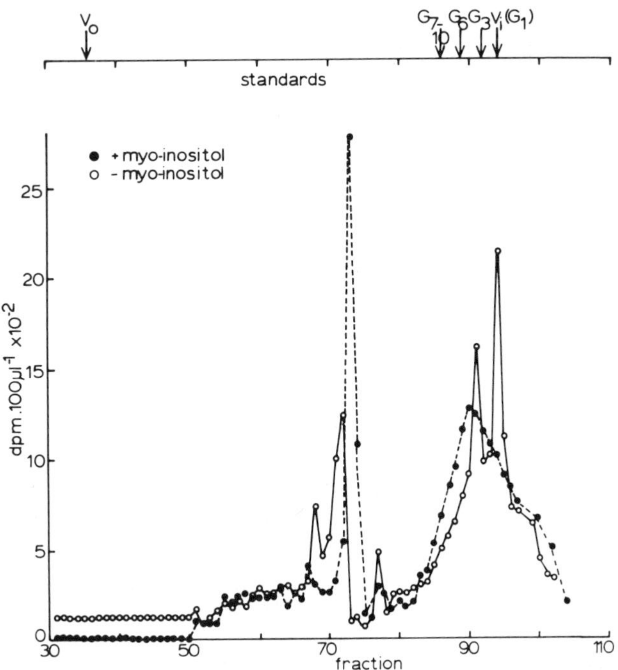

Fig. 6. Separation by Bio-Gel P-4 chromatography of the carbohydrates released by isolated endosperms after incubation in the presence (●) or absence (○) of 250 mM myo-inositol. Endosperms were dissected from dark-imbibed seeds after 4 h and incubated in a large volume for a further 26 h. Released sugars were then quantitatively radioisotope labeled *in vitro* using tritiated sodium borohydride,[20] before separation on a 1.5 x 98 cm Bio-Gel P-4 column (approximately 400 mesh; Bio-Rad) by elution with boiled deionized water. Aliquots (100 μl) were removed from each fraction and activity determined by liquid scintillation counting. Vo: void volume; Vi: inclusion volume; G_1, G_3, G_6, G_{7-10}: size of authentic glucose standards from monomer (G_1) to decamer (G_{10}).

Table 3. Effects of incubation volume on isolated lettuce endosperms.[a]

Incubation volume	α-Galactosidase activity	D-Galactose	β-Mannanase activity
ml	units	μg	units
2.0 (large)	7.1	6.6	8.6
0.1 (small)	7.9	<1.0	0.4

[a] After 24 h of incubation, batches of 20 endosperms were harvested for analysis. Results are expressed as per 20 endosperms. Assay procedures for α-galactosidase: Leung and Bewley;[14] for D-galactose: Schachter;[19] for endo-β-mannanase: Halmer and Bewley.[11]

tosidase is provided with substrate already made available by at least partial hydrolytic cleavage of the endosperm cell wall.

Cellulase and Hemicellulases

Because of the confining nature of the endosperm, and the fact that it imposes dormancy upon dark-imbibed seeds by restricting embryonic radicle emergence, the suggestion has been made that light (and hormones) overcome dormancy by causing the production and/or release of an enzyme to weaken the endosperm cell walls, and thus permit radicle elongation.[17] endo-β-Mannanase is clearly not a candidate for this role, for its production is exclusively post-germinative.[12] Nor can α-galactosidase be an effective dormancy-breaking enzyme because it cannot hydrolyze the cell wall until after cleavage of oligomers therefrom by mannanase. Cellulase has been suggested,[17] but this enzyme is present only in very small amounts in the dry seed (at the lowest limit of detection by viscometric assay) and it does not increase in activity above controls prior to germination of red- or gibberellin-induced seeds (Fig. 7). Its activity remains very low even during post-germinative endosperm hydrolysis. β-Mannosidase, α-galactosidase and xylanase are present in the dry seed but do not increase prior to or following germination of light-stimulated seeds. Nor do they change in activity in relation to the levels in dark-imbibed (dormant) seeds (data not presented). It is unlikely, therefore, that these three enzymes play any role in dormancy-breaking.

SUMMARY

The major polymeric sugar component of the cell walls of the lettuce seed endosperm is mannan, possibly galactomannan, which requires endo-β-mannanase and β-mannanase for hydrolysis, as well as α-galactosidase to reduce the size of mannanase-released oligomers. Hydrolysis is exclusively a post-germinative event, and neither cellulase nor mannanase (nor other hemicellulases) appears to play a role in dormancy breaking. That is, they do not appear to be produced in response to a germination stimulus prior to the completion of germination in order to weaken the constraining endosperm cell wall and thus permit radicle emergence.

endo-β-Mannanase is produced within the endosperm and is released into the cell wall (the endosperm is thus an autolytic tissue) in response to light- and hormone-induced stimuli initiated in the axis and transferred through the cotyledons to the endosperm. α-Galactosidase is present in the dry seed, and it increases in activity following a light

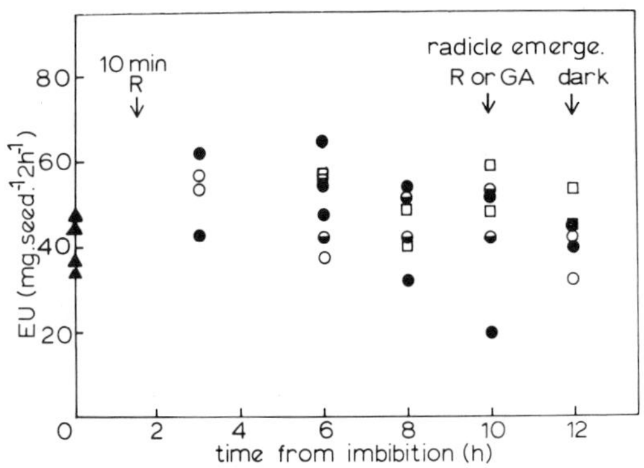

Fig. 7. Cellulase activity in dry (▲) intact lettuce seeds, in seeds imbibed in GA_3 (□), on water in darkness (●), or incubated on water in darkness but irradiated after 90 min with red light (O). Enzyme activity was extracted in 1 M NaCl and assayed viscometrically using carboxymethyl cellulose (Hercules Powder Co.) in a Cannon-Manning semi-microviscometer.[21]

or hormone stimulus. This enzyme cannot release galactose from the native cell wall, but it is important in reducing the size of oligomers released by prior mannanase activity. β-Mannosidase is not regulated by either light or hormones, but it is present within the cotyledons, apparently tightly associated with the cell walls, from the start of imbibition. It plays a role in mannan mobilization only when oligomers released by mannanase from the endosperm cell wall (and perhaps reduced further in size by α-galactosidase) diffuse to the cotyledons.

ACKNOWLEDGMENTS

This work is supported by the Natural Sciences and Engineering Research Council of Canada grant A6352. D.W.M.L. is the recipient of an Izaak Walton Killam Memorial Scholarship and F.B.O. is supported by teaching and research assistantships from the University of Calgary. J.D.B. wishes to thank Dr. G. A. Maclachlan for making available his laboratory for the cellulase experiments.

REFERENCES

1. BORTHWICK HA, WW ROBBINS 1928 Lettuce seed and its germination. Hilgardia 3: 275-305
2. JONES RL 1974 The structure of the lettuce endosperm. Planta 121: 133-146
3. BORTHWICK HA, SB HENDRICKS, EH TOOLE, VK TOOLE 1954 Action of light on lettuce seed germination. Botan Gaz 115: 205-225
4. IKUMA H, KV THIMANN 1964 Analysis of germination processes of lettuce seed by means of temperature and anaerobiosis. Plant Physiol 39: 756-767
5. LONA F 1956 L'acido gibberellico determina la germinazione dei semi di _Lactuca scariola_ in fase di skotoinhibizione. Ateneo Parmense 27: 641-646
6. KAHN A, JA GOSS, DE SMITH 1957 Effects of gibberellin on germination of lettuce seeds. Science 125: 645-646
7. BEWLEY JD, P HALMER 1980/1981 Embryo-endosperm interactions in the hydrolysis of lettuce seed reserves. Israel J Bot 29: 118-132
8. HALMER P, JD BEWLEY, TA THORPE 1978 Degradation of the endosperm cell walls of _Lactuca sativa_ L., cv Grand Rapids. Timing of mobilisation of soluble sugars, lipid and phytate. Planta 139: 1-8

9. LEUNG DWM, JSG REID, JD BEWLEY 1979 Degradation of the endosperm cell walls of Lactuca sativa L., cv Grand Rapids in relation to the mobilisation of proteins and the production of hydrolytic enzymes in the axis, cotyledons and endosperm. Planta 146: 335-341
10. HALMER P, JD BEWLEY, TA THORPE 1975 Enzyme to degrade lettuce endosperm cell wall during gibberellin- and light-induced germination. Nature (London) 258: 716-718
11. HALMER P, JD BEWLEY 1979 Mannanase production by the lettuce endosperm. Control by the embryo. Planta 144: 333-340
12. HALMER P, JD BEWLEY, TA THORPE 1976 An enzyme to degrade lettuce endosperm cell walls. Appearance of a mannanase following phytochrome- and gibberellin-induced germination. Planta 130: 189-196
13. LEUNG DWM, JD BEWLEY 1981 Immediate phytochrome action in inducting α-galactosidase activity in lettuce seeds. Nature (London) 289: 587-588
14. LEUNG DWM, JD BEWLEY 1981 Red-light and gibberellic-acid-enhanced α-galactosidase activity in germinating lettuce seeds, cv Grand Rapids. Control by the axis. Planta 152: 436-441
15. SHARMA CB 1971 Selective inhibition of α-galactosidases by myo-inositol. Biochem Biophys Res Commun 43: 572-579
16. REID JSG, JD BEWLEY 1979 A dual role for the endosperm and its galactomannan reserves in the germinative physiology of fenugreek (Trigonella foenum-graecum L.), an endospermic leguminous seed. Planta 147: 145-150
17. IKUMA H, KV THIMANN 1963 The role of seed-coats in germination of photosensitive lettuce seeds. Plant Cell Physiol 4: 169-185
18. REID JSG, H MEIER 1973 Enzymic activities and galactomannan mobilisation in germinating seeds of fenugreek (Trigonella foenum-graecum L. Leguminosae). Secretion of α-galactosidase and β-mannosidase by the aleurone layer. Planta 112: 301-308
19. SCHACHTER H 1975 Enzymic microassays for D-mannose, D-glucose, D-galactose, L-fucose and D-glucosamine. Methods Enzymol 41: 3-10
20. TAKASAKI S, A KOBATA 1974 Microdetermination of individual neutral and amino sugars and N-acetylneur-

aminic acid in complex saccharides. J Biochem 76: 783-789

21. BYRNE H, NV CHRISTOU, DPS VERMA, GA MACLACHLAN 1975 Purification and characterization of two cellulases from auxin-treated pea epicotyls. J Biol Chem 250: 1012-1018

Chapter Eight

TRANSPORT AND METABOLISM OF ASYMMETRICALLY-LABELED SUCROSE IN PEA EPICOTYLS

GORDON MACLACHLAN AND RANGIL SINGH*

Department of Biology
McGill University
1205 Avenue Docteur Penfield
Montreal, Quebec, Canada H3A 1B1

INTRODUCTION

During germination and growth of seedlings, sucrose is the major form in which carbon in general, and carbohydrate in particular, is transported from storage tissues through the seedling axis to growing regions. When starch is the major seed reserve, it is converted to sugar phosphates via action of phosphorylase or amylases and hexokinase, and sucrose is formed by sucrose or sucrose phosphate synthase.[1-6] When fructosans are major reserves, they are hydrolyzed during germination to oligosaccharides which rapidly convert to sucrose before translocation.[7] When lipid is the major reserve, it is converted via acetyl-CoA and reverse glycolysis to sugar phosphates and hence to sucrose.[8-13] When protein reserves are substantial, amino acids and amides may also be generated and transported from the seed,[14,15] but sucrose is

---

* Permanent address: Dept. of Biochem., Punjab Agric. Univ., Ludhiana, India.

still the major translocate. The only known exceptions to this generalization are those few plants, including cucumber, that transport galactosylated sucrose derivatives of the raffinose series.[16-22]

The question of what advantages accrue to plants as a result of the apparently universal preference for sucrose as translocate has been raised by many reviewers[23-25] and the alternatives discussed in specific detail.[26-27] It is also pertinent to ask why hexoses are not major constituents of transport streams[22,28] since most tissues other than vascular contain glucose (fructose in a few cases) as the predominant sugar.

In general it is supposed that sucrose is the transport sugar of choice because reducing groups of both hexose moieties are employed in the α-1,2 linkage, and therefore are unavailable to react with other metabolites enroute from source to sink. Reaction would require sucrose to be hydrolyzed (invertase) or cleaved (sucrose synthase reversal). A few indications have been found for active uptake or sucrose permease which would facilitate movement of sucrose from the transport system into other cells.[28a,b] Most of the evidence indicates that sucrose does not readily pass through plasmamembranes as such, but must first be converted to hexoses before uptake and utilization by non-vascular cells.[6,29-34]

Accordingly, sucrose may have an advantage over hexoses as a relatively unreactive transport form of sugar which is confined to the vascular system or apoplast until it encounters enzyme(s) that convert it to hexose. As such, the utilization of this form of carbon could be regulated by genetic and other (eg. hormonal) controls over the development and distribution of invertase and possibly sucrose synthase. Invertase activity, in particular, is known to be concentrated in growing regions of roots and stems.[35-42]

The possibility that the relatively high free energy of hydrolysis of sucrose (-6.6 kcal/mole) is useful to the plant is usually dismissed on the grounds that it represents at best a gain of 1.3% in potential metabolic free energy over that of invert sugars.[26] Nevertheless, it is conceivable that a strategically-located sucrose phosphorylase or synthase could generate hexose phosphate or UDP-glucose preferentially from sucrose which would then represent a relatively direct metabolic route to other useful products. Pea stem preparations have been shown to convert sucrose to β-glucans _in vitro_ via sucrose synthase (forming UDP-Glc) and membrane-bound β-glucan synthase.[43] The difficulty is that preferential utilization of one (or the other) hexose moiety of sucrose has never been shown to take place _in vivo_, and reaction _in vitro_ may represent an artifactual combination of enzyme activities that do not occur naturally in proximity.

The fact is that very little experimental evidence has been assembled to show that sucrose confers any particular advantage over hexose(s) as a result of its selection for transport. ^{14}C-labeled hexoses are rapidly converted to uniformly ^{14}C-labeled sucrose in all tissues that have been studied (e.g. Ref. 44, 45), and therefore sucrose is made available for transport whichever substrate is supplied. Sucrose labeled only in the glucose or fructose moiety has been employed in a number of such studies,[31,34,46-52] which generally show that supplied sucrose is, in part, hydrolyzed to hexoses which are incorporated and equilibrated in metabolic pools and recombined to form [U-^{14}C]sucrose.

In the present study the transport and metabolism of ^{14}C-labeled sugars has been investigated in etiolated pea stem tissues, either by supplying sugars to intact 8-day-old epicotyls cut just above the cotyledons, or to excised sections of tissue removed from apical growing and basal maturing regions. Particular use was made of asymmetrically ^{14}C-labeled sucrose to investigate the degree to which one hexose or the other was employed for synthesis of metabolites and end products. A preliminary account of part of this work has appeared[53] and related results are found elsewhere.[54]

GROWTH OF PEA EPICOTYLS IN SUCROSE VERSUS HEXOSES

When 8-day-old epicotyls were cut just above the cotyledons and immersed in nutrient solutions containing 2% sugar, they elongated more rapidly in sucrose than in 2% glucose or fructose, or 1% glucose plus 1% fructose (Fig. 1). The growth differential was evident within 2 days and continued for at least 8 days. Growth in fresh weight was also significantly greater when sucrose was supplied than hexose(s).[54] Using excised segments of tissue from elongating region of epicotyl, growth over a short period of time (6 h) was the same in glucose and/or fructose and in non-metabolized mannitol, but growth in length and fresh weight was significantly greater in sucrose (Table 1).

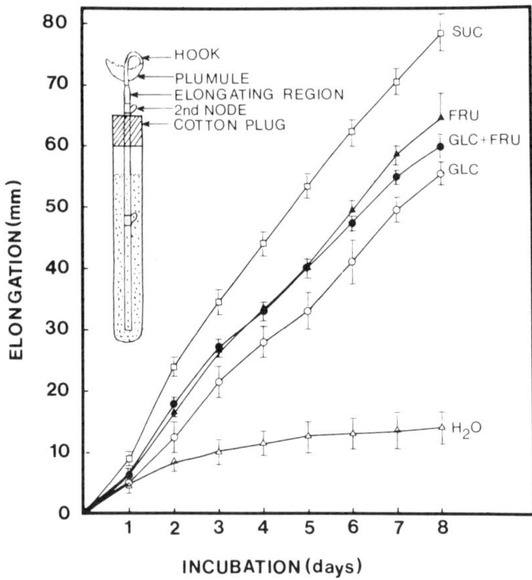

Fig. 1. Elongation of detached pea epicotyls supplied with sucrose or hexose(s). Seedlings of Pisum sativum var. Alaska were grown for 8 days in darkness, epicotyls were cut just above the cotyledons and immersed in sterile nutrient solution containing 2% (w/v) sucrose, glucose and/or fructose. The solution was maintained at 5°C to prevent microbial growth and the aerial parts (see diagram) were allowed to elongate at ambient temperature. Increase in length was measured at daily intervals. Values are given ± S.E. of 10 replicas.

Table 1. Growth of excised segments of elongating region of epicotyl[a] in sucrose versus hexose(s).

Medium	Elongation	Δ Fresh weight
	(mm/6 h ± S.E.)	(mg/seg)
2% Mannitol	0.6 ± 0.07	3.4
2% Glucose (Glc)	0.6 ± 0.08	3.3
2% Fructose (Fru)	0.6 ± 0.07	2.8
1% Glc + 1% Fru	0.8 ± 0.06	4.1
2% Sucrose (Suc)	1.1 ± 0.07	4.5

[a] Initial segments were 10 mm long, 10.6 mg fresh wt.

METABOLISM OF THE HEXOSE MOIETIES OF SUCROSE BY EXCISED PEA SEGMENTS

Segments (5 mm) were removed from different regions of pea epicotyls, washed thoroughly in distilled water and incubated in ([U-^{14}C]glc)sucrose or ([U-^{14}C]fru)sucrose (New England Nuclear). After 2 h at 30°C, up to one-half the sucrose in the external incubation medium was found to be hydrolyzed (Table 2). Hydrolysis was most extensive (30-50%) with segments from the elongating region and adjacent second node; it did not occur in solutions bathing the segments from the most apical or basal regions of epicotyl. It appears, therefore, that substantial invertase activity remains associated with cut cell surfaces of tissue excised from certain regions of epicotyl, despite repeated washing. This presumably reflects localized concentrations of invertase in cell walls.[30,34] One consequence, of course, is that some tissue segments were not presented with intact sucrose only in these tests, but with a mixture of sucrose and invert sugar. This point must be born in mind when interpreting data on metabolism.

Table 2. Invertase activity at cut surfaces of excised pea epicotyl segments floated on sucrose solution.[a]

Segment	Hydrolysis of sucrose in medium[b]	
	([U-^{14}C]glc)Suc	([U-^{14}C]fru)Suc
	%	
Plumule	0.0	0.0
Hook	15.1	11.0
Elongating region (0-5 mm)	31.8	51.7
Second node	42.4	49.5
Second internode	10.5	17.6
First node	5.1	0.0
First internode	0.0	0.0

[a] 10 segments (5 mm) in 250 µl 0.5% (w/v) ^{14}C-sucrose (1 µCi) at 30°C for 2 h.

[b] Determined by ratio of ^{14}C in hexose:sucrose following paper chromatography in n-butanol - glacial acetic acid - H_2O (4:1:5, upper layer).

The main sugar found endogenously in all regions of epicotyl was glucose (35-60% of total soluble saccharide) with sucrose never predominating (<15%) and fructose always a very minor component (<5%). A typical chromatogram showing the range of saccharides is given in Fig. 2. Sugar phosphates and nucleotides are barely detectable by this method but in

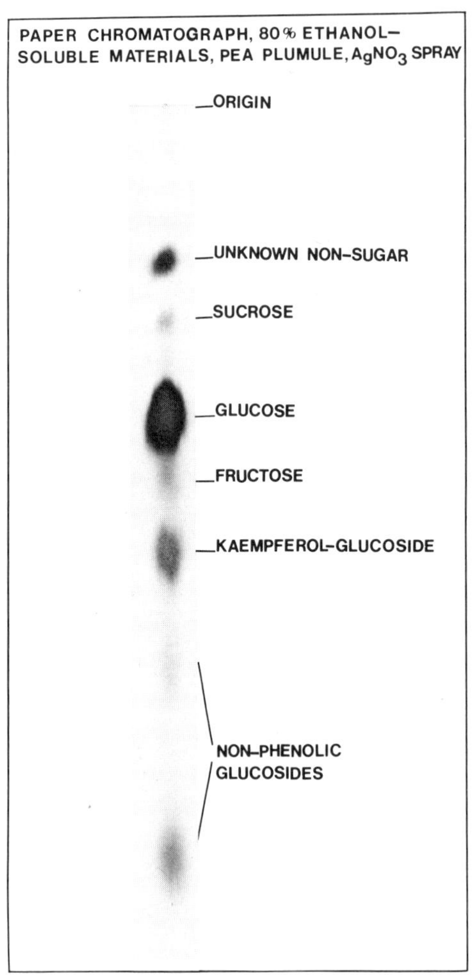

Fig. 2. Paper chromatograph of endogenous saccharides in ethanol-soluble extracts of pea plumules. The solvent system was n-butanol - acetic acid - H_2O (4:1:5, upper layer).

the most apical regions of epicotyl (plumule, hook) total concentrations of kaempferol glucoside and two non-phenolic glucosides were considerable (25-30%).

Chromatographs showed that label from the two ^{14}C-hexose moieties of supplied sucrose was incorporated into all of the sugars and glycosides found naturally in EtOH-soluble extracts (Fig. 2) except in excised segments from basal regions (Table 3). Here, very little free ^{14}C-hexose was formed, presumably because of low invertase levels (Table 2). In more apical regions, however, ^{14}C-sucrose was no longer the major labeled saccharide and ^{14}C was found in both hexoses, indicating that supplied sucrose had been hydrolyzed and the hexoses interconverted in metabolic pools. Complete or nearly complete equilibration between the hexoses was evident in the hook and plumule segments. Measurements of the ^{14}C distribution between hexose moieties of extracted sucrose indicated that most of the asymmetry of the supplied substrate was retained in all segments (>80%), including the plumule and hook.

Table 4 shows the relative amount of ^{14}C incorporated from asymmetrically-labeled sucrose into kaempferol glucoside

Table 3. Distribution of ^{14}C between sugars extracted from excised segments supplied with asymmetrically-labeled ^{14}C-sucrose.[a]

Segment	Total ethanol-soluble ^{14}C					
	([U-^{14}C]glc)Suc			([U-^{14}C]fru)Suc		
	Suc	Glc	Fru	Suc	Glc	Fru
	%					
Plumule	23	22	23	24	16	24
Hook	19	24	20	18	19	31
Elongating region	34	51	7	34	12	47
Second node	40	52	5	41	9	48
Second internode	80	23	2	73	3	24
First node	89	9	2	94	1	5
First internode	95	3	2	96	1	2

[a] Segments were incubated as described in Table 2.

Table 4. Incorporation of ^{14}C from asymmetrically-labeled sucrose into kaempferol-glucoside and ethanol-insoluble materials by excised pea epicotyl segments.[a]

Segment	Total recovered ^{14}C in:			
	Kaempferol-glucoside		Ethanol-insoluble products	
	([U-^{14}C]glc)Suc	([U-^{14}C]fru)Suc	([U-^{14}C]glc)Suc	([U-^{14}C]fru)Suc
		%		
Plumule	13.2	11.6	20.3	19.4
Hook	17.1	15.0	14.6	13.4
Elongating region	3.3	4.4	4.3	4.1
Second node	not detectable		4.8	4.5
Second internode	not detectable		2.5	1.8
First node	not detectable		1.5	1.2
First internode	not detectable		0.7	0.5

[a] Segments were incubated as described in Table 2.

and EtOH-insoluble products by various excised segments. The insoluble materials do not include starch[54] but products that hydrolyze to a mixture of ^{14}C-sugars typical of cell wall polysaccharides.[55] The results indicate that neither hexose moiety of sucrose was preferentially used to synthesize soluble glucoside or cell wall polysaccharide in any region of epicotyl. Clearly, these products must have been synthesized from a metabolic pool where the hexoses from sucrose were equilibrated, not directly from sucrose by transglucosylation via UDP-glucose (sucrose synthase).

TRANSPORT AND METABOLISM OF SUGARS BY DETACHED INTACT PEA EPICOTYLS

The bases of detached etiolated epicotyls (8-day-old, with 2 nodes and developing third internode) were supplied with equal amounts of ^{14}C in 0.5% (w/v) sucrose, glucose or fructose. After 2 h about 6% of the label had been taken up into the epicotyl in all cases. Fig. 3 shows radioautographs of chromatographs of EtOH-soluble materials extracted from various regions of the detached epicotyls. The following conclusions are evident from the radioautographs: a) In basal regions (first and second internodes and first node) ^{14}C-sucrose was the main soluble product whether the supplied ^{14}C-sugar was hexose or asymmetrically-labeled sucrose. b) Whichever hexose was ^{14}C-labeled in the substrate provided, that particular hexose tended to predominate in the pool of free ^{14}C-hexoses found throughout the epicotyl. c) In apical regions (plumule, hook, elongating region), ^{14}C appears in all of the major soluble saccharides which occur naturally in the epicotyl (c.f. Fig. 2), whichever substrate was provided. d) More ^{14}C appeared in the plumule when ^{14}C hexoses were provided at the base rather than ^{14}C sucrose.

In addition, quantitative tests[54] showed that the sucrose extracted from the various segments when ^{14}C-hexoses were supplied was uniformly labeled in both hexose moieties up to the elongating region, but there and above, the extracted sucrose was more labeled in the hexose moiety that had been supplied at the base. In contrast, when asymmetrically-labeled sucrose was supplied, asymmetry was maintained up to the elongating region, but sucrose from the hook and plumule contained ^{14}C-hexoses that were completely equilibrated. Thus, supplied ^{14}C-sucrose moved up the epicotyl as such only as far as the elongating region. There it was completely hydrolyzed before ^{14}C moved into more apical regions. Supplied ^{14}C-hexoses also translocated as such, at least in part, through the epicotyl to accumulate in the second node and at the apex at a rate faster than the translocation of ^{14}C from sucrose.

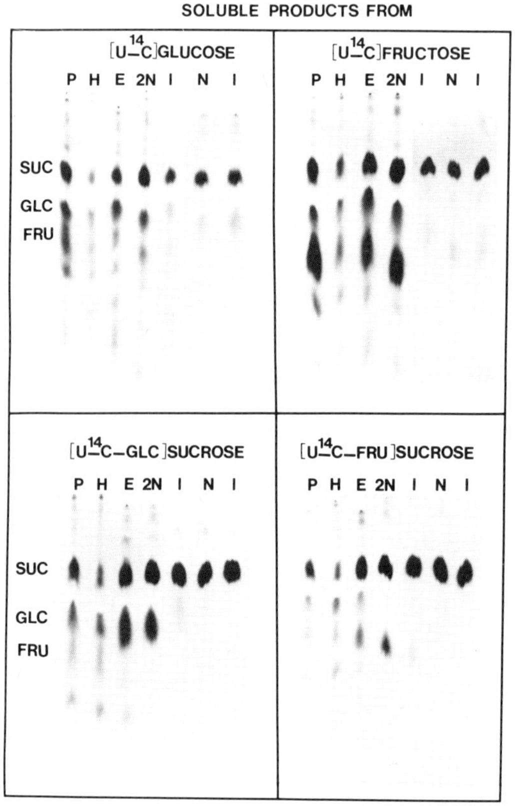

Fig. 3. Radioautographs of chromatographs of ethanol-soluble materials in various epicotyl regions following uptake of ^{14}C-sugars through cut bases of intact epicotyls. Detached epicotyls were supplied for 2 h with 200 μl 0.5% ^{14}C-sugars (1 μCi/epicotyl), segments were removed (letters refer to plumule (P), hook (H), elongating region (E), etc., as in Table 2) and extracted with hot 80% ethanol. Aliquots were chromatographed (as in Fig. 2) and radioautographed (Dupont-Cronex-4 X-ray film). Kaempferol glucoside is the main labeled product with mobility just below fructose (see Fig. 2) that is concentrated in apical regions.

THE "INVERTASE BARRIER" TO SUCROSE TRANSPORT

After grinding pea tissues in a mortar and removing insoluble material by centrifugation (x 50,000 g), essentially all of the invertase activity detected between pH 4 and 8 remained in the supernatant.[56] It was assayed reductometrically[57] using buffer-soluble enzyme from different regions of the epicotyl and 0.1 M sucrose as substrate. One unit of activity was defined as the amount of enzyme required to catalyze the production of 10 μg hexose/h at 30°C at the pH optimum, 5.0.

Table 5 shows the distribution of invertase activity in the plumule, hook and successive 10 mm segments of the third internode of the epicotyl. Activity levels were highest in the elongating region (0-10 mm) on a segment or fresh weight (volume) basis, and in the region just basal to this on a soluble protein basis. Activities in the first and second internodes were much lower. Thus total invertase activity, as well as that at cut surfaces (cf. Table 2), is highly localized in apical tissues of etiolated pea seedlings which are actively or just completing their elongation, in contrast to numerous other hydrolases, eg. β-glucosidase, amylase, etc., which are most concentrated at the extreme apex (plumule).[37]

Several other lines of evidence besides the restricted distribution of invertase may be cited to support the view

Table 5. Distribution of total invertase activity in plumule, hook and third internode of 8-day-old pea epicotyl.

Region	Invertase activity[a]		
	Per seg	Per 10 mg fresh wt	Per mg protein
	units of activity		
Plumule	3	2	4
Hook	3	8	29
0-10 mm from hook	79	36	283
10-20 mm from hook	49	19	448
20-30 mm from hook	26	10	311

[a] Details are given in text and in Maclachlan et al.[37]

that the levels of this enzyme are closely regulated in pea seedlings. Decapitation of epicotyls, and particularly excision of the elongating region, resulted in a marked reduction in the capacity for growth and a concurrent decay in the invertase levels (Table 6). Invertase activity was barely detectable in the elongating region within three days after decapitation (Table 7), and one day after excision (Table 6). The loss of invertase activity was not due to the development of an invertase inhibitor or protease activity, since addition of pea extracts from excised tissues with low invertase activity to extracts of tissues with high invertase

Table 6. Decay of invertase levels in elongating region of pea epicotyls and protection by indoleacetic acid (IAA).[a]

Treatment	Changes in elongating region after one day's growth		
	Length	Fresh wt.	Invertase units/seg
	Ratio: $\frac{\text{Value after 1 day's treatment}}{\text{Initial value in 8 day epicotyls}}$		
Intact epicotyl	2.2	2.8	1.1
Excised	1.2	1.4	0.03
Decapitated	1.2	1.5	0.4
+ Gibberellic acid	1.9	2.5	0.3
+ IAA	1.2	2.2	23.9
+ IAA + Actinomycin D	1.3	1.8	6.5
+ IAA + Cycloheximide	1.2	1.8	6.8

[a] After Datko.[56] Changes in the elongating region (apical 10 mm of third internode) of 8-day-old epicotyls were measured either in the undisturbed seedling (intact epicotyl), in the segment of tissue removed from the seedling (excised) and floated on buffer, or in the tissue left attached to the seedling but with plumule and hook removed (decapitated). In the latter case, effects of added hormones, etc., (10 μg/apex) were tested by applying these in lanolin paste.

levels did not result in any activity losses.[56] The decay *in vivo* could be delayed and even reversed for a time in decapitated epicotyls by the application of increasing levels of indoleacetic acid (Table 7). Effects of this growth-promoting hormone were not duplicated by others, eg. gibberellic acid, and they appear to have been dependent on RNA and protein synthesis since inhibitors of the latter processes not only prevented the growth response to hormone but also the protection of invertase levels (Table 6).

Thus pea seedlings, like other stem and root apices, appear to generate invertase only in regions which are in the process of cell elongation. Activity is maintained as long as growth continues, after which the enzyme rapidly disappears. Some of this enzyme may be secreted into extracellular compartments since the distribution of invertase activity at cut tissue surfaces (Table 2) resembles that extracted from the tissue (Table 5). Accordingly, any sucrose that is translocated from basal regions of the epicotyl would be expected to encounter high invertase activity in the elongating region only, regardless of the compartment(s) in which the sugar was transported. It is in this sense that it is possible to consider the sharp localization of invertase in elongating regions as a regulated barrier to the transport of sucrose into more apical regions, and to explain why intact sucrose is transported only from the base up to that point.

Table 7. Indoleacetic acid treatment delays the decay of invertase levels in decapitated pea epicotyls.[a]

Time	Invertase activity in elongating region			
	-IAA control	+IAA 0.1 μg	+IAA 1.0 μg	+IAA 10 μg
days	(units/apical 10 mm segment)			
0		141		
1	28	103	118	236
2	3	31	52	79
3	1	19	14	9

[a] After Datko.[56]

CONCLUSIONS

Several possible explanations for why cut pea epicotyls and segments grow better with supplied sucrose than hexose(s) (Table 1, Fig. 1) can be eliminated at once.

a) Carbon derived from supplied sucrose does not travel from the cut base of the epicotyl to the growing regions more rapidly than that from supplied hexose(s); on the contrary, in short-term experiments (30 min), hexose carbon accumulates in the plumule to levels 4 to 5x greater than that from sucrose.[54] Accordingly, sucrose confers no quantitative nutritional advantage over hexose(s) as translocate to the growing apex.

b) No specific metabolite or end product could be detected that was formed from supplied sucrose but not from hexoses, or vice-versa, in any part of the epicotyl (Fig. 3). Nor were any products that might be considered particularly useful for growth, such as wall polysaccharides or kaempferol glucoside, generated more rapidly from the glucose moiety of sucrose than from the fructose moiety (Table 4). Thus, in pea seedlings sucrose synthase does not appear to operate *in vivo* in a direction leading to sucrose cleavage and UDP-glucose production.

c) Regions of cell division in the epicotyl (hook) cannot be responsible for the enhanced growth observed in sucrose since sucrose supplied at the cut base never reaches this region as such. Rather, it moves up to the elongating region intact and there it is inverted, incorporated into metabolic pools and the hexose moieties completely interconverted.[54] Thus, whichever moiety of supplied sucrose is ^{14}C-labeled, the sucrose extracted from the plumule and hook is uniformly-labeled and therefore newly synthesized from a hexose pool. In addition, of course, it is directly demonstrable that segments of elongating regions which do not carry out cell division nevertheless show a positive growth response to sucrose over hexose(s) (Table 1).

d) Basal regions of the epicotyl have no difficulty in synthesizing sucrose from supplied hexoses (Fig. 3), and presumably newly-formed sucrose is transported as effectively as supplied sucrose. Not only is sucrose synthase activity distributed throughout the epicotyl[37] but it is clearly active *in vivo*. Supplied hexoses must be part of the translocate also, since they accumulate at the apex as such (Fig. 3). The presence of more of one labeled hexose over the other in apical regions could not be explained by hydrolysis of translocated sucrose, since the latter was equally labeled in both hexose moieties whichever hexose was supplied.[54]

These results focus attention on the distribution and regulation of invertase activity, since it appears that translocated sucrose cannot be metabolized by pea tissues without first being hydrolyzed. In basal regions of the epicotyl, extracellular invertase activity is not present (Table 2) and it is barely detectable in extracts of the tissue. Thus supplied sucrose would be expected to translocate through this region intact (Table 3) and not be utilized to any significant extent until it encountered the invertase activity barrier in elongating regions (Table 2 and Table 5). In fact, the basal regions barely metabolize supplied sucrose into any other products (Fig. 3, Table 2 and Table 4), and the invertase activity in elongating regions is clearly effective *in vivo* in rendering supplied sucrose available for metabolism. The activity in this region is under tight regulatory control by the plant, for it is not only subject to rapid turnover but its levels depend on supply of the auxin type of hormone (Table 6 and Table 7).

It may be concluded therefore, that pea epictoyls supplied with free hexoses are able to use them directly for metabolism in elongating regions, but those supplied with sucrose are limited in ability to metabolize it by the distribution and activity of invertase. This explains why there is a greater flux of hexoses into metabolic pools in elongating regions when hexoses are supplied than when sucrose is supplied.[54] This could have consequences for differential growth quite apart from the quantitative differences in metabolism. It is possible, for example, that extracellular spaces and cell walls of elongating tissue maintain relatively more acid conditions when they are supplied with sucrose than hexose as a result of the existence of a proportional hexose/proton symport.[58] This in turn could lead to better growth in sucrose (or suppressed growth in hexoses) in so far as extracellular acidity appears to be necessary for rapid cell expansion.[59] This possibility is at least amenable to experimental testing and may be one direction that research in this area should take in the future.

ACKNOWLEDGMENT

Financial support by grants (to G.M.) from the Natural Sciences and Engineering Research Council of Canada and le Programme de Formation de Chercheurs et d'Action Concertée du Québec is gratefully acknowledged.

REFERENCES

1. EDELMAN J, SI SHIBKO, AJ KEYS 1959 The role of the scutellum of cereal seedlings in the synthesis and transport of sucrose. J Exp Bot 10: 178-189
2. HAWKER, JS 1971 Enzymes concerned with sucrose synthesis and transformations in seeds of maize, broad bean and castor bean. Phytochemistry 10: 2313-2322
3. JULIANO BO, JE VARNER 1969 Enzymic degradation of starch granules in cotyledons of germinating peas. Plant Physiol 44: 886-892
4. KEYS AJ, SJ SKEWS 1961 Sucrose synthesis in the living scutellum of wheat seedlings. Biochem J 79: 13p
5. NOMURA T, T AKAZAWA 1973 Enzymic mechanism of starch breakdown in germinating rice seeds. IV. De novo synthesis of sucrose 6-phosphate synthetase in scutellum. Plant Physiol 51: 979-981
6. ap REES T 1974 Pathways of carbohydrate breakdown in higher plants. In DH Northcote, ed, MTP International Review of Science, Biochemistry Series 1, Vol 11, Plant Biochemistry, Butterworths, London pp 89-127
7. BHATIA IS, SK MANN, R SINGH 1974 Biochemical changes in the water-soluble carbohydrates during the development of chicory (Cichorium intybus) roots. J Sci Food Agric 25: 535-539
8. BREIDENBACH RW, H BEEVERS 1967 Association of the glyoxylate cycle enzymes in a novel subcellular particle from castor bean endosperm. Biochem Biophys Res Commun 27: 462-469
9. CANVIN DT, H BEEVERS 1961 Sucrose synthesis from acetate in the germinating castor bean: kinetics and pathway. J Biol Chem 236: 988-995
10. HEYDEMAN MT 1958 Isocitritase in germinating marrow seeds. Nature, London, 181: 627-628
11. KORNBERG HL, H BEEVERS 1957 The glyoxylate cycle as a stage in the conversion of fat to carbohydrate in the castor beans. Biochim Biophys Acta 26: 531-537

12. MARCUS A, J VELASCO 1960 Enzymes of glyoxylate cycle in germinating peanuts and castor beans. J Biol Chem 235: 563-567
13. STUMPF PK, C BRADBEER 1959 Fat metabolism in higher plants. Annu Rev Plant Physiol 10: 197-222
14. LARSON LA, H BEEVERS 1965 Amino acid metabolism in young pea seedlings. Plant Physiol 40: 424-432
15. STEWART CR, H BEEVERS 1967 Gluconeogenesis from amino acids in germinating castor bean endosperm and its role in transport to embryo. Plant Physiol 42: 1587-1595
16. HENDRIX JE 1968 Labeling pattern of translocated stachyose in squash. Plant Physiol 43: 1631-1636
17. HENDRIX JE 1982 Sugar translocation in two members of the Cucurbitaceae. Plant Sci Lett 25: 1-7
18. PRISTUPA NA 1959 The transport form of carbohydrates in pumpkin plants. Fiziol Rast 6: 30-35
19. WEBB KL, JWA BURLEY 1964 Stachyose translocation in plants. Plant Physiol 39: 973-977
20. WEBB JA, PR GORHAM 1964 Translocation of photosynthetically assimilated ^{14}C in straight-necked squash. Plant Physiol 39: 663-672
21. ZIMMERMANN MH 1957 Translocation of organic substances in trees. I. The nature of the sugars in the sieve tube exudate of trees. Plant Physiol 32: 288-291
22. ZIEGLER H 1975 Nature of transported substances. In MH Zimmermann, JA Milburn, eds, Transport in Plants. I. Phloem Transport, Encyclopedia of Plant Physiology, New Series, Vol 1, Springer-Verlag, Berlin pp 59-100
23. AKAZAWA T, K OKAMOTO 1980 Biosynthesis and metabolism of sucrose. In J Preiss, ed, The Biochemistry of Plants, Vol 3, Academic Press, New York pp 199-220
24. GIAQUINTA RT 1980 Translocation of sucrose and oligosaccharides. In J Preiss, ed, The Biochemistry of Plants, Vol 3, Academic Press, New York pp 271-320
25. PONTIS HG 1977 Riddle of sucrose. In DH Northcote, ed, International Review of Biochemistry, Plant Biochemistry II, Vol 13, Univ. Park Press, Baltimore, Maryland pp 79-117
26. ARNOLD WN 1968 The selection of sucrose as the translocate of higher plants. J Theor Biol 21: 13-20
27. PONTIS HG 1978 On the scent of the riddle of sucrose. Trends Biol Sci 3: 137-139
28. ESCHRICH W, W HEYSER 1975 Biochemistry of phloem constituents. In MH Zimmermann, JA Milburn, eds, Transport in Plants. I. Phloem Transport, Encyclopedia

of Plant Physiology, New Series, Vol 1, Springer-Verlag, Berlin, pp 101-136

28a. DESHUSSES J, SC GUMBER, FA LOEWUS 1981 Sugar uptake in lily pollen. A proton symport. Plant Physiol 67: 793-796

28b. THORNE JH 1982 Characterization of the active sucrose transport system of immature soybean embryos. Plant Physiol 70: 953-958

29. ESCHRICH W 1980 Free space invertase, its possible role in phloem unloading. Ber Dtsch Bot Ges 93: 363-378

30. GLASZIOU KT 1961 Accumulation and transformation of sugars in stalks of sugarcane. Origin of glucose and fructose in the inner space. Plant Physiol 36: 175-179

31. HAWKER JS, MD HATCH 1965 Mechanism of sugar storage by mature stem tissue of sugarcane. Physiol Plant 18: 444-453

32. HO LC 1980 Control of import into tomato fruits. Ber Dtsch Bot Ges 93: 315-325

33. HUMPHREYS TE 1974 Sucrose transport and hexose release in the maize scutellum. Phytochemistry 13: 2387-2396

34. SACHER JA, MD HATCH, KT GLASZIOU 1963 Sugar accumulation cycle in sugarcane. III. Physical and metabolic aspects of cycle in immature storage tissue. Plant Physiol 38: 348-354

35. HELLEBUST JA, DF FORWARD 1962 The invertase of the corn radicle and its activity in successive stages of growth. Can J Bot 40: 113-126

36. LYNE RL, T ap REES 1971 Invertase and sugar content during differentiation of roots of _Pisum sativum_. Phytochemistry 10: 2593-2599

37. MACLACHLAN GA, AH DATKO, J ROLLIT, E STOKES 1970 Sugar levels in the pea epicotyl: Regulation by invertase and sucrose synthetase. Phytochemistry 9: 1023-1030

38. MACLEOD RD, D FRANCIS 1977 Some observations on invertase activity in roots of _Vicia faba_ L. J Exp Bot 28: 853-863

39. RICARDO CPP, T ap REES 1970 Invertase activity during the development of carrot roots. Phytochemistry 9: 239-247

40. ROBINSON E, R BROWN 1952 The development of the enzyme complement in growing root cells. J Exp Bot 3: 356-374

41. SEITZ K, A LANG 1968 Invertase activity and cell growth in lentil epicotyls. Plant Physiol 43: 1075-1082

42. SEXTON R, SUTCLIFFE JF 1969 The distribution of β-glycerophosphatase in young roots of _Pisum sativum_ L. Ann Bot 33: 407-419

43. ROLLIT J, GA MACLACHLAN 1974 Synthesis of wall glucan from sucrose by enzyme preparations from Pisum sativum. Phytochemistry 13: 367-374
44. KRIEDEMANN P, H BEEVERS 1967 Sugar uptake and translocation in the castor bean seedling. II. Sugar transformations during uptake. Plant Physiol 42: 174-180
45. TAMMES PML, CR VONK, J VAN DIE 1973 Studies on phloem exudation from Yucca flaccida Haw. XI. Xylem feeding of ^{14}C-sugars and some other compounds, their conversion and recovery from the phloem exudate. Acta Botan Neerl 22: 233-237
46. DICK PS, T ap REES 1975 The pathway of sugar transport in roots of Pisum sativum. J Exp Bot 26: 305-314
47. GIAQUINTA RT 1977 Phloem loading of sucrose. pH dependence and selectivity. Plant Physiol 59: 750-755
48. GIAQUINTA RT 1977 Sucrose hydrolysis in relation to phloem translocation in Beta vulgaris. Plant Physiol 60: 339-343
49. GIAQUINTA RT, W LIN, N SADLER, V FRANCESCHI 1982 Symplastic pathway of sucrose unloading from the phloem. Plant Physiol 69(suppl): 5
50. HATCH MD, JA SACHER, KT GLASZIOU 1963 Sugar accumulation cycle in sugarcane. I. Studies in enzymes of the cycle. Plant Physiol 38: 338-343
51. HATCH MD, KT GLASZIOU 1964 Direct evidence for translocation of sucrose in sugarcane leaves and stems. Plant Physiol 39: 180-184
52. PORTER HK, LH MAY 1955 Metabolism of radioactive sugars by tobacco leaf discs. J Exp Bot 6: 43-63
53. SINGH R, GA MACLACHLAN 1982 Pea epicotyls grow better in sucrose than in hexose(s). Plant Physiol 69(suppl): 148
54. SINGH R, GA MACLACHLAN 1983 Transport and metabolism of sucrose vs. hexoses in relation to growth in etiolated pea stem. Plant Physiol 71: in press
55. MACLACHLAN GA, CT DUDA 1965 Changes in concentration of polymeric components in excised pea-epicotyl tissue during grwoth. Biochim Biophys Acta 97: 288-299
56. DATKO AH 1968 Development of glycoside hydrolase and pectic enzyme activities in growing pea epicotyl tissue. Ph.D. Thesis, McGill University, Montreal
57. HODGE JE, BT HOFREITER 1962 Determination of reducing sugars and carbohydrates. In RL Whistler, ML Wolfrom, eds, Methods in Carbohydrate Chemistry, Vol 1, Academic Press, New York pp 380-394

58. KOMOR E, M THOM, A MARETZKI 1981 The mechanism of sugar uptake by sugar cane suspension cells. Planta 153: 181-192

59. RAYLE DL, R CLELAND 1977 Control of plant cell enlargement by hydrogen ions. In AA Moscona, A Monroy, eds, Current Topics in Developmental Biology, Vol 11, Academic Press, New York pp 187-214

Chapter Nine

PHYTATE METABOLISM WITH SPECIAL REFERENCE TO ITS *MYO*-INOSITOL COMPONENT

FRANK A. LOEWUS

Institute of Biological Chemistry and
Program in Biochemistry/Biophysics
Washington State University
Pullman, Washington 99164-6340

INTRODUCTION

To most plant scientists, the term phytic acid suggests that compound most commonly functioning as the major phosphate reserve of the seed. Indeed, the carbon skeleton of phytic acid is dismissed as a structure of small consequence to the seedling.[1,2] Phytate *is* a major source of inorganic phosphate (P_i) for the germinated seed.[3-5] In this respect its importance ranks that of other seed reserves such as carbohydrate, fat and protein, compounds dependent upon P_i for mobilization during germination. It is also an ester of the alicyclic polyol, *myo*-inositol (MI). Structurally, phytic acid is the hexakisphosphoric acid ester of MI (Fig. 1). All six phosphates are *ortho*[6] and the fully dissociated salt has twelve negative charges. The sodium salt hydrate containing 38 molecules of water has an inverted conformation with axial phosphates at carbon 1, 3, 4, 5 and 6 and an equatorial phosphate at carbon 2[7] as shown in Figure 2. In solution, conformation is influenced by pH with the structures in Figure 2 representing conformational extremes.[8-10] Within the seed, where phytate occurs as a complex salt of K, Mg and

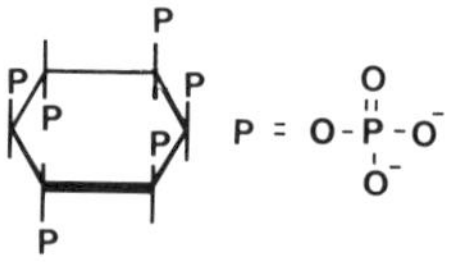

Fig. 1. Haworth structure of myo-inositol hexakisphosphate (phytic acid). Short lines to carbon represent hydrogen atoms.

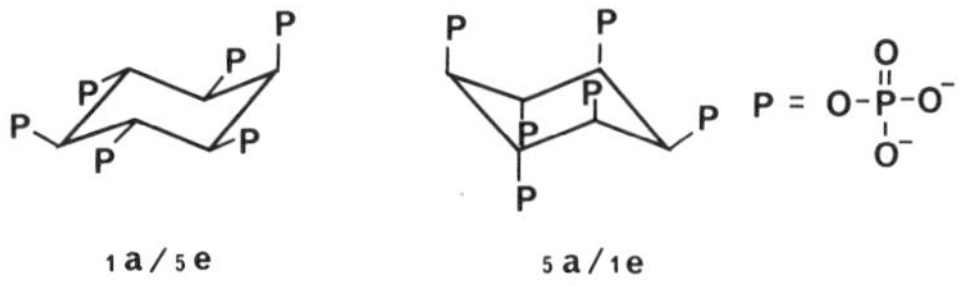

Fig. 2. Conformational structures for myo-inositol hexakisphosphate in chair forms. Hydrogen substituents to carbon are deleted.

in some instances Ca,[11-13] its conformation is still unknown. Partial salt-type linkages between phytate and protein in phytate-rich regions of the seed are not excluded.

Analysis of phytic acid by high performance liquid chromatography and application of that method to the measurement of phytate levels in foodstuffs was recently reported.[13a,b] Phytic acid interactions in food systems was recently reviewed by Cheryan[13c] and by Reddy et al.[13d]

During seed maturation, phytate is deposited in discrete regions (globoids) of organelles which for descriptive purposes are generally assigned to the class of subcellular particles referred to as protein bodies.[12,14,15] Formation and accumulation of phytate is restricted to such subcellular regions.[16-18] Its subsequent utilization as a source of P_i also occurs at these sites.[19] This immobility of phytate, a quality similar to that of starch and other reserve polysaccharides of the seed, invites consideration of biosynthetic processes which lead to formation and accumulation of phytate on one hand and to redistribution of products of phytate breakdown on the other. Here the biosynthesis and metabolism of MI play significant roles. This chapter examines these aspects with special emphasis on the MI component of phytate.

BIOSYNTHESIS OF *MYO*-INOSITOL MONOPHOSPHATE AND PHYTATE

From the viewpoint of seed development, D-glucose-6-P is the precursor of the MI requirement for phytate biosynthesis since 1L-MI-1-P synthase (Fig. 3) is the sole *de novo* mechanism for MI formation in plants, animals and MI-synthesizing microorganisms.[20] The chiral form of the product from lily pollen synthase is 1L (Fig. 4), identical to that produced by synthase from animal tissue or yeast.[21] Presumably, MI-1-P synthase from other plant sources also produces 1L-MI-1-P although this must still be tested. A direct role for synthase-produced 1L-MI-1-P in phytate biosynthesis has been proposed[22,23] but experimental evidence is lacking. Others regard MI-2-P (Fig. 5) as the MI-monophosphate precursor of phytate,[24,25] the latter drawing their conclusion from *in vivo* labeling studies and from identification of an *in vitro* product of phosphorylation of MI as MI-2-P. These assignments must be treated with caution since acid-catalyzed migration of phosphate between carbon 1 or 3 and carbon 2 is possible.[26] Moreover, an enzymatic interconversion between 1L-MI-1-P and MI-2-P which involves a cyclic MI-1,2-monophosphate is also conceivable.

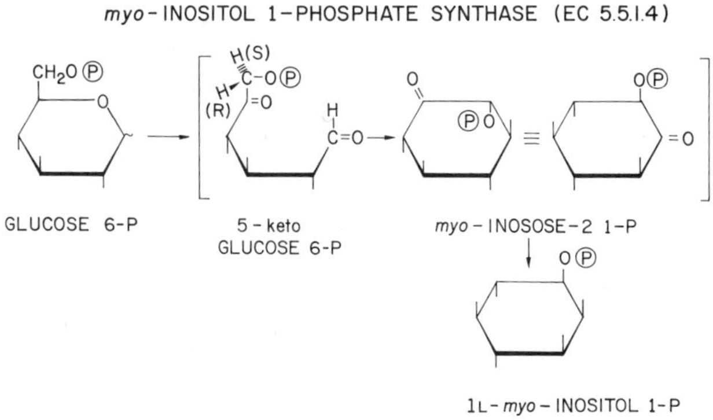

myo-INOSITOL 1-PHOSPHATE SYNTHASE (EC 5.5.1.4)

GLUCOSE 6-P

5-keto GLUCOSE 6-P

myo-INOSOSE-2 1-P

1L-*myo*-INOSITOL 1-P

Fig. 3. Proposed mechanism of *myo*-inositol 1-phosphate synthase. Phosphate is given by the symbol Ⓟ. The prochiral configuration of hydrogen at carbon 6 of 5-keto glucose-6-P is illustrated. Cofactor (not shown) for oxidation of glucose-6-P to 5-keto-glucose-6-P and for reduction of *myo*-inosose-2 1-P to *myo*-inositol-1-P is NAD. Reproduced with permission from the Annual Review of Plant Physiology, Volume 34. © 1983 by Annual Reviews, Inc.

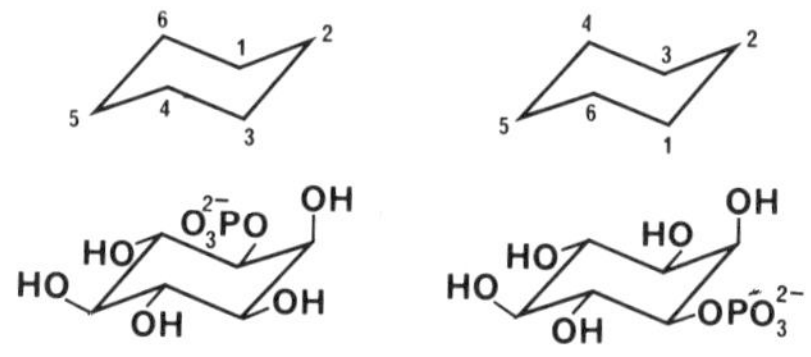

Fig. 4. Chiral forms of *myo*-inositol-1-phosphate. The rule for numbering separate carbon atoms is illustrated above each structure. Hydrogen substituents to carbon are deleted.

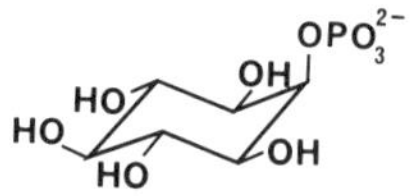

Fig. 5. Structure of *myo*-inositol-2-phosphate.

MI-1-P is hydrolyzed to free MI by a heat-stable, Mg^{2+}-dependent alkaline phosphatase.[27] This enzyme exhibits an affinity about 10-fold greater for 1D- and 1L-MI-1-P than for MI-2-P. In plants free MI participates in many metabolic processes, notably the MI oxidation pathway, galactinol biosynthesis, phosphatidylinositol biosynthesis and the formation of isomeric inositols and their ethers.[28,29] It also serves as substrate for MI kinase (Fig. 6), an enzyme widely distributed in nature.[30,31] MI kinase, like MI-1-P synthase, produces 1L-MI-1-P[21] and in this respect provides the plant with a salvage mechanism for recycling free MI into the 1L-MI-1-P pool. If phosphorylation of 1L-MI-1-P rather than MI-2-P is regarded as the first committed step in phytate biosynthesis, then two mechanisms exist for supplying precursor, the synthase which relies on hexose phosphate production and the kinase which is dependent on a renewable source of free MI. Plants and plant tissues readily convert free MI to phytate.[32-34] In a comparative study on the efficiency of labeling of phytate in maturing seed of *Sinapis alba* by D-[U-^{14}C]glucose and [U-^{14}C]MI, it was found that MI was several times more efficient[34] but both substrates were useful in this regard.

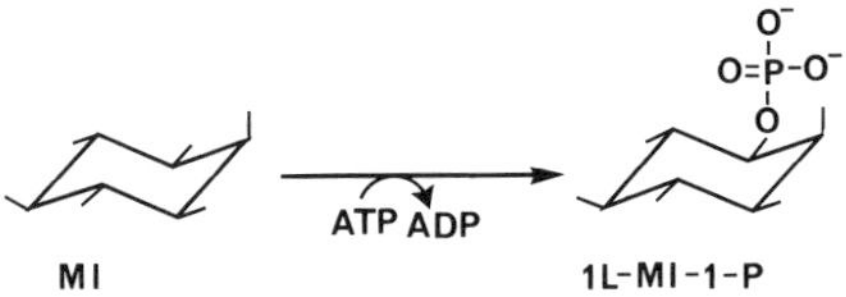

Fig. 6. The reaction catalyzed by myo-inositol kinase.

For subsequent phosphorylation of MI monophosphate to phytate, a single enzyme is proposed, phosphoinositol kinase.[35,36] This enzyme catalyzes a stepwise ATP-Mg^{2+} dependent phosphorylation of MI monophosphate through a phosphoprotein intermediate.[37] Curiously, germinating mung bean seed rather than maturing seed provided the source of this enzyme and there are no reports of the isolation of phosphoinositol kinase from ripening seeds or grain in which phytate synthesis is proceeding at a rapid rate. In as much as phytate accumulates in discrete regions within phytate-synthesizing subcellular particles, it is these particles in which the phosphoinositol kinase should be found. Related processes which provide the ATP requirement and supply the counter ion in the globoid still require biochemical interpretations.[4]

HYDROLYSIS OF PHYTATE AND RELEASE OF FREE MYO-INOSITOL

During seed germination, the phytate-rich organelles are slowly depleted of phytate, a process dependent on the presence of phytase which is produced during germination. Wheat bran phytase, the most completely studied enzyme of this class,[4,38,39] consists of a 6-phytase and possibly, a 2-phytase. The prefixed numeral refers to the position of the first phosphate to be removed. In the case of 6-phytase, the first product is 1L-MI-1,2,3,4,5-pentaphosphate and subsequent steps remove phosphate from carbons 1 (or 4 or 5), 5 (or 4), 4, and 3 to produce MI-2-P, the ultimate product. Action of 2-phytase[38] gives MI-1-P, possibly the chiral 1L form although this remains to be proved. Other phytases from bacterial and fungal sources exhibit other specificities.[4] In each case, the ultimate product is MI monophosphate and a second enzyme, a MI monophosphate phosphatase, is required to generate free MI and release the final P_i. One such enzyme has been isolated from lily pollen but its affinity for MI-2-P (K_m = 0.73 mM) is an order of magnitude less than it is for MI-1-P.[27] An acid phosphatase present in phytate-rich organelles of rice grains may also play a role in hydrolyzing MI-monophosphate that has been generated during germination.[40]

Hormonal control, similar to that involved in starch hydrolysis, may determine phytase formation and secretion during germination. Studies on isolated aleurone layers and gibberellic acid-treated half-seeds of barley indicate that phytate hydrolysis requires this hormone to effect P_i release.[41,42] Since hydrolysis proceeds beyond MI monophosphate to P_i, the effect of treatment with gibberellic acid must also extend to the phosphatase required for hydrolysis of MI monophosphate.

Phytase activity appears to be associated with the phytate-rich organelle.[19] How the hormonal signal is transmitted to this site and then expressed are questions still to be answered.

A scheme based on currently available information for processes involved in the formation and breakdown of phytate is given in Figure 7.

THE MYO-INOSITOL OXIDATION PATHWAY

The central role of D-glucose-6-P in the metabolism of carbohydrates in plants is well established.[43] Conversion of glucose-6-P into nucleotide sugars fulfills biosynthetic needs of the growing plant and supplies the sugar components which are required for transport, storage and development.

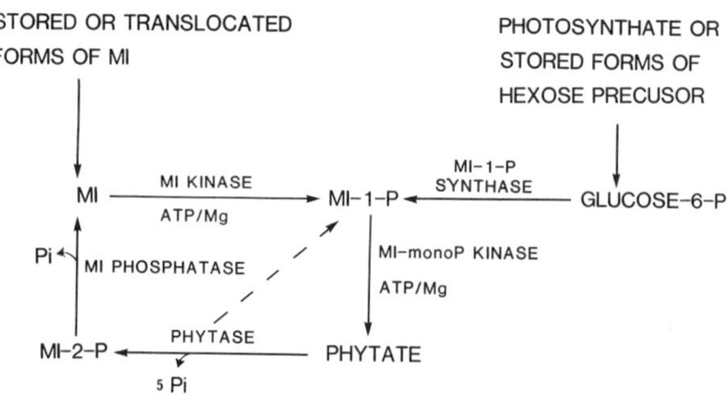

Fig. 7. A scheme depicting the biosynthesis and breakdown of *myo*-inositol hexakisphosphate (phytic acid). *myo*-Inositol is abbreviated MI.

One very important nucleotide sugar intermediate is uridine diphosphate D-glucuronate (UDPGlcA). Not only does it provide glucuronosyl units during polysaccharide biosynthesis, it also functions as the precursor of other glycosyl units which are needed during polysaccharide production, namely UDP-D-galacturonate (UDPGalA), UDP-D-xylose (UDPXyl), UDP-L-arabinose (UDPAra) and UDP-D-apiose (UDPApi). These nucleotide sugars with the possible exception of UDPApi are virtually ubiquitous in the plant kingdom and their sugar moieties are encountered in non-cellulosic cell wall polysaccharides, plant glycoproteins, gums and mucilages.[44]

Plants contain two enzymatic pathways from glucose-6-P to UDPGlcA and its metabolic products (Fig. 8). In the nucleotide sugar oxidation pathway on the right-hand side of this scheme, the key enzyme is UDPGlc dehydrogenase which was isolated from pea seedlings.[45] This enzyme oxidizes carbon 6 of UDPGlc in two discrete 2-electron NAD^+-linked steps. Both the uridylyltransferase that forms UDPGlc and the dehydrogenase are product inhibited. Moreover, nucleotide sugar products of UDPGlcA, notably UDPXyl, also inhibit the dehydrogenase. The requirement for NAD^+ and the kinetic restraints imposed by product inhibition are important considerations

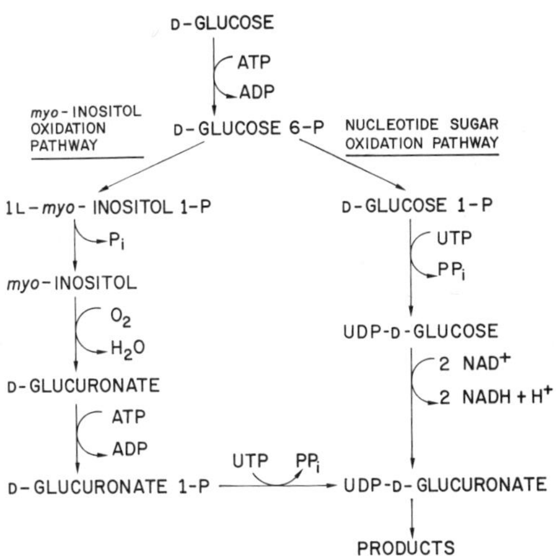

Fig. 8. Alternative pathways from D-glucose-6-P to UDP-D-glucuronate biosynthesis in plants. Reproduced with permission from the Annual Review of Plant Physiology, Volume 34. © 1983 by Annual Reviews, Inc.

when invoking the nucleotide sugar oxidation pathway as a route to UDPGlcA.

The left-hand scheme, conveniently referred to as the MI oxidation pathway, was first proposed in 1962.[46] It consists of five steps; 1L-MI-1-P synthase, MI-1-P phosphatase, MI oxygenase, D-glucuronokinase and D-glucuronate-1-P uridylyl-transferase.[46a] One may consider the first two reactions as a biosynthetic phase since both 1L-MI-1-P and free MI are substrates for other metabolic requirements as well as the MI oxidation pathway, and free MI, supplied exogenously or translocated within the plant from a site other than that involved in polysaccharide biosynthesis, is effective as substrate for this pathway. The final three steps are largely committed to UDPGlcA production. Carbon flowing through the MI oxidation pathway to UDPGlcA occurs without a coupled reduction of NAD^+ as is the case in the other pathway and bypasses centrally positioned UDPGlc for which there is a highly competitive demand due to its biosynthetic role.[47]

A number of experimental approaches were taken to weigh the relative functionality of the two pathways.[48,49] In germinating lily pollen, strong evidence was obtained which favored a major functional role for the MI oxidation pathway during pectin biosynthesis in elongating pollen tube walls. Experimentally, this study involved comparison of ^3H:^{14}C ratios in glucosyl and galacturonosyl residues of tube wall polysaccharide from germinated lily pollen that had been labeled with D-[5-^3H,1-^{14}C]glucose. Galacturonosyl residue derived from the MI oxidation pathway would be expected to show the large carbon 5 hydrogen isotope effect created by 1L-MI-1-P synthase.[50] The effect was found in galacturonate from tube wall pectin. Had the other pathway prevailed, such an effect would not be observed since oxidation of the glucose precursor involves carbon 6 instead of carbon 5. At the time it was difficult to reconcile MI-1-P synthase activity with galacturonate synthesis, that is, the amount of MI produced by synthase was insufficient to account for pectin synthesis even if the free MI reserve of the resting pollen grain was taken into consideration.[51] A very recent development offers a solution to this dilemma. Certain pollens, among them pollen from *Lilium* species, contain appreciable phytate.[52,52a] In *L. henryi*, the phytate content is 1.1% by weight, possibly sufficient to supply the increment of free MI that synthase activity failed to produce. Thus *Lilium* pollen, which has been the experimental tissue of choice for studies on the MI oxidation pathway,[27,48,53-56] may now serve a similar role in studies of phytate biosynthesis and metabolism.

MOBILIZATION OF RESERVES WITH REGARD TO PHYTATE IN GERMINATING WHEAT

Small grains as typified by wheat offer an excellent opportunity to trace the events which on one hand lead to accumulation of phytate and on the other, to mobilization of this reserve during germination. The appearance of Type I inclusions or phytin globoids in aleurone cells of the wheat caryopsis is morphologically distinguishable about 14 days after anthesis.[57] If, at this time, wheat plants are supplied with [2-^{3}H]MI (Fig. 9) by injection of a solution of the label into the hollow lumen of the peduncle, about 50% of the label is translocated to the spike.[18] The portion not translocated is recovered in the region of the puncture created by the injection and is largely utilized as substrate for the MI oxidation pathway into polysaccharide formation in the wounded region of the peduncle. Virtually all of the label that is translocated into the spike is recovered in the developing kernels, especially the bran fraction which includes the aleurone layer. When labeled MI is replaced by scyllo-[R-^{3}H]inositol ([R-^{3}H]SI) (Fig. 9) as the source of label for injection, none of the label is metabolically trapped in the region of injection since SI is not a substrate for the MI oxidation pathway and almost all of the label accumulates in the spike.

Successive extraction of the bran fraction with 10% dimethyl sulfoxide (25°C, 6 h), 0.2% Na_2EDTA (50°C, 1 h) and 2 M trifluoroacetic acid (100°C, 2 h) gave results listed in Table 1. From [2-^{3}H]MI-labeled bran, dimethyl sulfoxide extracted three labeled components, free MI, galactinol and

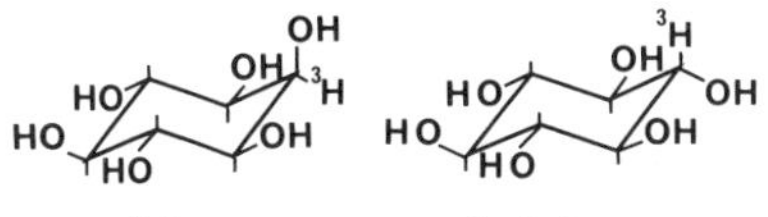

Fig. 9. Tritium-labeled structures of myo-[2-^{3}H]inositol and scyllo-[R-^{3}H]inositol. Protium atoms are indicated by short lines. The term R refers to the random position of tritium in all-trans scyllo-inositol.

Table 1. Distribution of tritium in wheat bran fractions from grain of *myo*-[2-^{3}H]inositol- and *scyllo*-[R-^{3}H]inositol-labeled wheat plants. Plants were labeled 2 weeks ([2-^{3}H]MI) or 4 weeks ([R-^{3}H]SI) post-anthesis and grown to maturity.

Fraction	Radioactivity	
	[2-^{3}H]MI	[R-^{3}H]SI
	%	
Soluble in 10% dimethyl sulfoxide	11	96
Soluble in 0.2% EDTA	41	4
Soluble in 2 M trifluoroacetic acid	38	<1
Insoluble in 2 M trifluoroacetic acid	5	0
Total Recovery	95	100

an unknown compound with chromatographic properties on cellulose intermediate to the first two compounds (Fig. 10). Virtually all of the labeled EDTA extractable material was phytic acid (Fig. 11). The trifluoroacetic acid soluble and insoluble fractions consisted of labeled products associated with the MI oxidation pathway, primarily arabinose and xylose which are major components of aleurone cell walls.[58] When [R-^{3}H]SI-labeled bran was extracted with 10% dimethyl sulfoxide, 96% of the tritium was solubilized. Cellulose column chromatography (Fig. 12) revealed two labeled components, free SI (peak I) and SI-galactoside (peak II).

These results demonstrate the highly directed utilization of translocable MI for phytate biosynthesis in developing wheat caryopses. Accompanying incorporation of label from [2-^{3}H]MI into aleurone layer cell wall polysaccharide may be due to intermediates of the MI oxidation pathway that were generated at the injection site and then translocated into the spike, or to utilization of a portion of the [2-^{3}H]MI afte it reached the caryopsis. Absence of labeled phytate in caryopses from [R-^{3}H]SI-labeled plants is a clear indication that this all *trans* isomer is not utilized in phytate biosynthesis even though it does follow the same translocatory path as MI and functions as an acceptor for the galactinol-synthesizing enzyme.

Wheat kernels from [2-^{3}H]MI- and [R-^{3}H]SI-labeled plants similar to those just described were used to follow the mobilization of reserves during germination.[59] Galactinol, which accounted for 58% of the 80% ethanol-soluble label in the embryo + scutellum of [2-^{3}H]MI-labeled grain was quickly hydrolyzed. Twenty hours after imbibition, only 2% remained. Free MI increased 5-fold in this tissue during the same period. A similar but less pronounced decrease in SI-galacto-

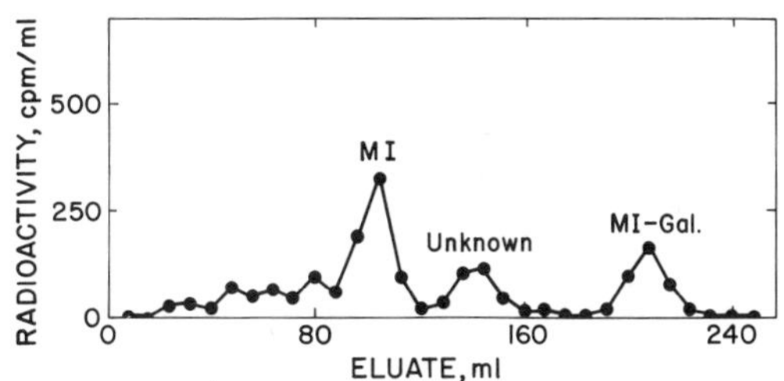

Fig. 10. Cellulose column chromatography of low molecular weight compounds that were extracted from [2-^{3}H]MI-labeled wheat bran with 80% ethanol. The eluent was acetone-water, 4:1. Abbreviations: myo-inositol, MI; galactinol, MI-gal. Reproduced from Ref. 18 with publisher's permission.

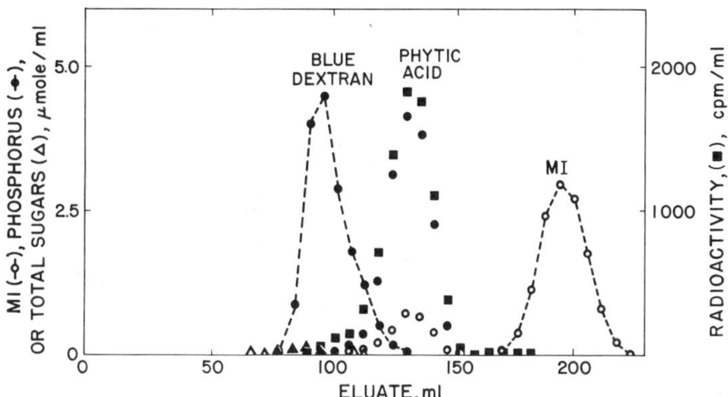

Fig. 11. Sephadex G-25 gel filtration of the EDTA extracted fraction from [2-^{3}H]MI-labeled wheat bran. Reproduced from Ref. 18 with publisher's permission.

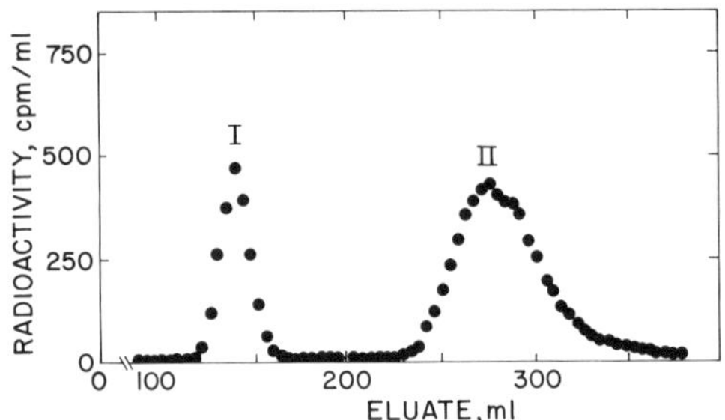

Fig. 12. Cellulose column chromatography of low molecular weight compounds from the dimethyl sulfoxide extract of [R-^{3}H]SI-labeled wheat bran after removal of high molecular weight polysaccharides by precipitation with three volumes of ethanol. Peak I is labeled scyllo-inositol and peak II is a mixture of six compounds of which only one, scyllo-inositol galactoside, is labeled. Reproduced from Ref. 18 with publisher's permission.

side accompanied by a rise in free SI was found in embryo + scutellum tissue from [R-^{3}H]SI-labeled grain.

As roots and shoot developed, labeled reserves were translocated into seedling parts (Fig. 13). Loss of tritium due to exchange with water or to leakage was minimal. By the third day, postgermination, all water-soluble phytate had been hydrolyzed but a more resistant form of phytate (solubilized only after treatment with EDTA and a heating step followed by treatment with commercial amyloglucosidase, which also contained pentosanase activity) followed a much slower course and continued to release MI for several days (Fig. 14). In this figure, the 80% ethanol-soluble fraction, mainly free MI, reflects the slow release of MI from this resistant form of phytate in the kernel.

During early stages of radicle emergence, free MI and galactinol-derived MI were the principal sources of labeled reserve, later augmented by breakdown of water-soluble phytate. Beyond 3 days, the labeling pattern in the seedling was complicated by the arrival of labeled breakdown products, primarily pentoses, from polysaccharidic material in the bran layers.[60] Although the dry weight of root tissue failed to increase beyond 7 days, conversion of tritium from translocated reserves into cell wall polysaccharides of the roots

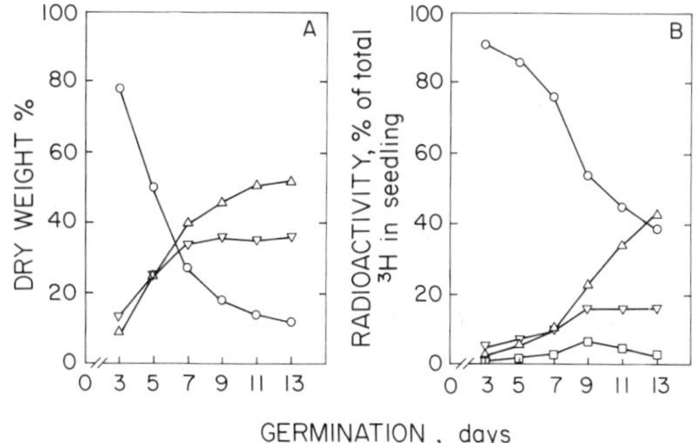

Fig. 13. Changes in dry weight and distribution of tritium in germinating grain from [2-^{3}H]MI-labeled wheat plants. A, dry weight; B, radioactivity. Symbols: (○) kernel, (▽) roots, (△) shoot, (□) radioactivity recovered from supporting filter paper. Reproduced from Ref. 59 with publisher's permission.

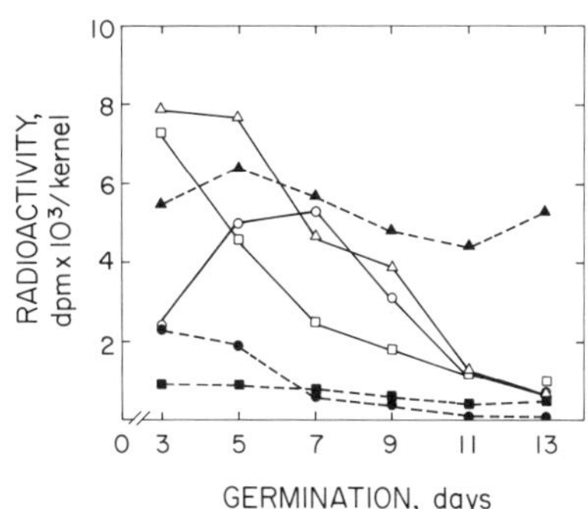

Fig. 14. Changes in the distribution of tritium in the residual kernel (less roots and shoot) of germinating grain from [2-^{3}H]MI-labeled wheat plants. Symbols for various fractions: (○) 80% ethanol-soluble, (●) EDTA-soluble, (△) water-soluble, (□) amyloglucosidase-soluble, (▲) trifluoroacetic acid-soluble and (■) trifluoroacetic acid-insoluble, extracted in that order. Reproduced from Ref. 59 with publisher's permission.

continued up to the end of the experimental period, ultimately accounting for 80% of the total label in the roots (Fig. 15). Shoot tissue accumulated more label than the roots and conversion of translocated MI into cell wall polysaccharides was the major pathway. During the 13 day period of germination, about 70% of the available label was transferred from reserves to the new seedling, largely to form cell wall polysaccharides.

When similar experiments were made with [R-^{2}H]SI-labeled grains, free SI moved from the kernel into the developing seedling where it remained as free SI, unable to function as substrate for the MI oxidation pathway.

These studies clearly demonstrate that phytate, in addition to its role as a phosphate reserve in seeds and grains, performs a second role as a source of MI. This phytate-derived MI along with stored reserves of free MI and galactinol-derived MI, supply the new seedling with substrate for the MI oxidation pathway and, ultimately, cell wall polysaccharides.

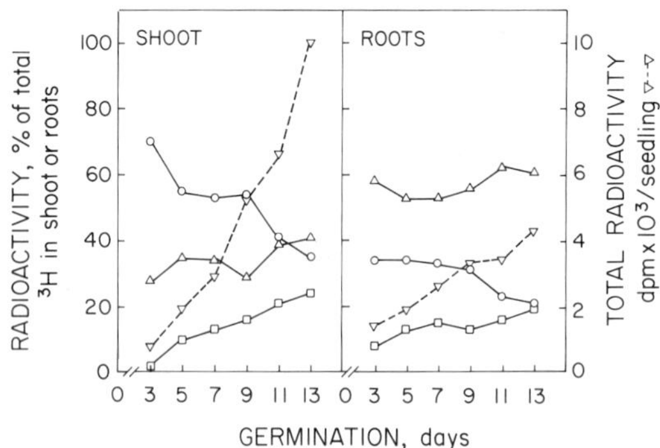

Fig. 15. Changes in the distribution of tritium in shoot and roots of seedlings from [2-^{3}H]MI-labeled wheat plants. Symbols: (△) 80% ethanol-soluble fraction, (○) trifluoroacetic acid-soluble fraction, (□) trifluoroacetic acid-insoluble fraction and (▽) total radioactivity per seedling. Reproduced from Ref. 59 with publisher's permission.

CONCLUDING REMARKS

Our understanding of phytate metabolism is still fragmentary. Biosynthetic events leading to MI monophosphate have been described but the relationship between this putative substrate and phytate is still obscure. Virtually nothing is known about phytate deposition. Events triggered by germination that lead to phytate breakdown and mobilization of the products of hydrolysis lag far behind our understanding of similar processes involving starch reserves. If this chapter does no more than to bring to light our lack of knowledge in this area, the effort has been worthwhile.

ACKNOWLEDGMENT

Work described herein was supported by a grant GM22427 from the National Institutes of Health and project 0266, College of Agriculture Research Center, Washington State University, Pullman, Washington.

REFERENCES

1. MAYER AM, Y SHAIN 1974 Control of seed germination. Annu Rev Plant Physiol 25: 167-193
2. MAYER AM, I MARBACH 1981 Biochemistry of the transition from resting to germinating state in seeds. Prog Phytochem 7: 95-136
3. BIELESKI RL 1973 Phosphate pools, phosphate transport, and phosphate availability. Annu Rev Plant Physiol 24: 225-252
4. COSGROVE DJ 1980 Inositol Phosphates: Their Chemistry, Biochemistry and Physiology. Elsevier, Amsterdam, Oxford, New York 191 p
5. MAGA JA 1982 Phytate: its chemistry, occurrence, food interactions, nutritional significance and methods of analysis. J Agric Food Chem 30: 1-9
6. JOHNSON LF, ME TATE 1969 Structure of "phytic acids." Can J Chem 47: 63-73
7. BLANK CE, J PLETCHER, M SAX 1971 The structure of myo-inositol hexaphosphate dodecasodium salt octatriacontahydrate: a single crystal X-ray analysis. Biochem Biophys Res Commun 44: 319-325
8. BROWN DM, ME TATE 1973 pH and cation-dependent conformation of cyclitol phosphates. Proc Aust Biochem Soc 6: 44-49

9. COSTELLO AJR, T GLONEK, TC MYERS 1976 ^{31}P Nuclear magnetic resonance-pH titrations of myo-inositol hexaphosphate. Carbohydr Res 46: 159-171
10. ZUIDERWEG ERP, GGM Van BEEK, SH DeBRUIN 1979 The influence of electrostatic interaction on the proton-binding behaviour of myo-inositol hexakisphosphate. Eur J Biochem 94: 297-306
11. OGAWA M, K TANAKA, Z KASAI 1979 Energy-dispersive X-ray analysis of phytin globoids in aleurone particles of developing rice grains. Soil Sci Plant Nutr 25: 437-448
12. LOTT JNA 1980 Protein bodies. In NE Tolbert, ed, The Biochemistry of Plants, Vol 1, The Plant Cell, Academic, New York pp 589-623
13. LOTT JN, JS GREENWOOD, CM VOLLMER 1982 Mineral reserves of castor beans: the dry seed. Plant Physiol 69: 829-833
13a. GRAF E, FR DINTZIS 1982 HPLC method for the determination of phytate. Anal Biochem 119: 413-417
13b. GRAF E, FR DINTZIS 1982 Determination of phytic acid in foods by high-performance liquid chromatography. J Agric Food Chem 30: 1094-1098
13c. CHERYAN M 1980 Phytic acid interactions in food systems. CRC Crit Rev Food Sci Nut 13: 297-335
13d. REDDY NR, SK SATHE, DK SALUNKHE 1982 Phytates in legumes and cereals. Adv Food Res 28: 1-92
14. MATILE P 1978 Biochemistry and function of vacuoles. Annu Rev Plant Physiol 29: 193-213
15. PERNOLLET J-C 1978 Protein bodies of seeds: ultrastructure, biochemistry, biosynthesis and degradation. Phytochemistry 17: 1473-1480
16. OGAWA M, K TANAKA, Z KASAI 1979 Phytic acid formation in dissected ripening rice grains. Agric Biol Chem 43: 2211-2213
17. OGAWA M, K TANAKA, Z KASAI 1979 Accumulation of phosphorus, magnesium and potassium in developing rice grains: followed by electron microprobe X-ray analysis focusing on the aleurone layer. Plant Cell Physiol 20: 19-27
18. SASAKI K, FA LOEWUS 1980 Metabolism of myo-[2-^{3}H]inositol and scyllo-[R-^{3}H]inositol in ripening wheat kernels. Plant Physiol 66: 740-745
19. YOSHIDA T, K TANAKA, Z KASAI 1975 Phytase activity associated with isolated aleurone particles of rice grains. Agr Biol Chem 39: 289-290

20. SHERMAN WR, MW LOEWUS, MZ PIÑA, Y-HH WONG 1981 Studies on myo-inositol-1-phosphate synthase from Lilium longiflorum pollen, Neurospora crassa and bovine testis. Further evidence that a classical aldolase step is not involved. Biochim Biophys Acta 660: 299-305
21. LOEWUS MW, K SASAKI, AL LEAVITT, L MUNSELL, WR SHERMAN, FA LOEWUS 1982 The enantiomeric form of myo-inositol-1-phosphate produced by myo-inositol-1-phosphate synthase and myo-inositol kinase in higher plants. Plant Physiol 70: 1661-1663
22. MOLINARI E, O HOFFMANN-OSTENHOF 1968 Untersuchungen über die biosynthese der cyclite. XXI Über ein enzymsystem das myo-inosit zu phytinsäure phosphorylieren kann. Hoppe-Syler's Z Physiol Chem 349: 1797-1799
23. DE BP, BB BISWAS 1979 Evidence for the existence of a novel enzyme system. myo-Inositol-1-phosphate dehydrogenase in Phaseolus aureus. J Biol Chem 254: 8717-8719
24. TANAKA K, K WATANABE, K ASADA, Z KASAI 1971 Occurrence of myo-inositol monophosphate and its role in ripening rice grains. Agric Biol Chem 35: 314-320
25 IGAUE I, M SHIMIZU, S MIYAUCHI 1980 Formation of a series of myo-inositol phosphates during growth of rice plant cells in suspension culture. Plant Cell Physiol 21: 351-356
26. POSTERNAK T 1965 The Cyclitols. Holden Day, San Francisco, Hermann, Paris
27. LOEWUS MW, FA LOEWUS 1982 myo-Inositol-1-phosphatase from the pollen of Lilium longiflorum Thunb. Plant Physiol 70: 765-770
28. LOEWUS FA, MW LOEWUS 1980 myo-Inositol: biosynthesis and metabolism. In J. Preiss, ed, The Biochemistry of Plants, Vol 3, Carbohydrates: Structure and Function, Academic, New York pp 43-76
29. LOEWUS FA, DB DICKINSON 1982 Cyclitols. In FA Loewus, W Tanner, eds, Encyclopedia of Plant Physiology, New Series, Vol 13A, Plant Carbohydrates I, Intracellular Carbohydrates. Springer Verlag, Berlin, Heidelberg, New York pp 193-216
30. ENGLISH PD, M DEITZ, P ALBERSHEIM 1966 Myoinositol kinase: partial purification and identification of product. Science 151: 198-199
31. TANAKA K, T YOSHIDA, Z KASAI 1976 Phosphorylation of myo-inositol by isolated aleurone particles of rice. Agr Biol Chem 40: 1319-1325

32. ROBERTS RM, FA LOEWUS 1968 Inositol metabolism in plants. VI Conversion of myo-inositol to phytic acid in Wolffiella floridana. Plant Physiol 43: 1710-1716
33. TANAKA K, T YOSHIDA, Z KASAI 1974 Radioautographic demonstration of the accumulation site of phytic acid in rice and wheat grains. Plant Cell Physiol 15: 147-151
34. BLAICHE FM, KD MUKHERJEE 1981 Radiolabeled phytic acid from maturing seeds of Sinapus alba. Z Naturforsch 36C: 383-384
35. MAJUMDER AL, NC MANDEL BB BISWAS 1972 Phosphoinositol kinase from germinating mung bean seeds. Phytochemistry 11: 503-508
36. CHAKRABARTI S, AL MAJUMDER 1978 Phosphoinositol kinase from plant and avian sources. In F. Eisenberg Jr, WW Wells, eds, Cyclitols and Phosphoinositides, Academic Press, New York pp 69-81
37. MAJUMDER AL, BB BISWAS 1973 Further characterization of phosphoinositol kinase isolated from germinating mung bean seeds. Phytochemistry 12: 315-319
38. TOMLINSON RV, CE BALLOU 1962 Myoinositol polyphosphate intermediates in the dephosphorylation of phytic acid by phytase. Biochemistry 1: 166-171
39. LIM PE, ME TATE 1973 The phytases. II Properties of phytase fractions F_1 and F_2 from wheat bran and the myo-inositol phosphates produced by fraction F_2. Biochim Biophys Acta 302: 316-328
40. YAMAGATA H, K TANAKA, Z KASAI 1979 Isoenzymes of acid phosphatase in aleurone particles of rice grains and their interconversion. Agr Biol Chem 43: 2059-2066
41. CLUTTERBUCK VJ, DE BRIGGS 1974 Phosphate mobilization in grains of Hordeum distichon. Phytochemistry 13: 45-54
42. KATAYAMA N, H SUZUKI 1980 Possible effect of gibberelli on phytate degradation in germinating barley seeds. Plant Cell Physiol 21: 115-123
43. GANDER JE 1982 Polyhydroxy acids: relation to hexose phosphate metabolism. In FA Loewus, W Tanner, eds, Encyclopedia of Plant Physiology, New Series, Vol 13A, Plant Carbohydrates I, Intracellular Carbohydrates, Springer-Verlag, Berlin, Heidelberg, New York pp 77-102
44. FEINGOLD DS 1982 Aldo (and keto) hexoses and uronic acids. In FA Loewus, W Tanner, eds, Encyclopedia of Plant Physiology, New Series, Vol 13A, Plant Carbohydrates I, Intracellular Carbohydrates, Springer Verlag, Berlin, Heidelberg, New York pp 3-76

45. STROMINGER JL, LW MAPSON 1957 Uridine diphosphoglucose dehydrogenase of pea seedlings. Biochem J 66: 567-572
46. LOEWUS FA, S KELLY, EF NEUFELD 1962 Metabolism of myo-inositol in plants: conversion to pectin, hemi-cellulose, D-xylose and sugar acids. Proc Nat Acad Sci USA 48: 421-425
46a. LOEWUS FA, MW LOEWUS 1983 myo-Inositol: its biosynthesis and metabolism. Annu Rev Plant Physiol 34: 137-161
47. FEINGOLD DS, G AVIGAD 1980 Sugar nucleotide transformations in plants. In J Preiss, ed, The Biochemistry of Plants, Vol 3, Carbohydrates: Structure and Function, Academic, New York pp 101-170
48. LOEWUS FA, MW LOEWUS, IB MAITI, C-L ROSENFIELD 1978 Aspects of myo-inositol metabolism and biosynthesis in higher plants. In F. Eisenberg Jr, WW Wells, eds, Cyclitols and Phosphoinositides, Academic, New York pp 249-267
49. LOEWUS MW, FA LOEWUS 1980 The C-5 hydrogen isotope-effect in myo-inositol 1-phosphate synthase as evidence for the myo-inositol oxidation-pathway. Carbohydr Res 82: 333-342
50. LOEWUS MW 1977 Hydrogen isotope effects in the cyclization of D-glucose 6-phosphate by myo-inositol-1-phosphate synthase. J Biol Chem 252: 7221-7223
51. MAITI IB, C-L ROSENFIELD, FA LOEWUS 1978 myo-Inositol content of lily pollen. Phytochemistry 17: 1185-1186
52. JACKSON JF, G JONES, HF LINSKENS 1982 Phytic acid in pollen. Phytochemistry 21: 1255-1258
52a. JACKSON JF, HF LINSKENS 1982 Phytic acid in Petunia hybrida pollen is hydrolysed during germination by a phytase. Acta Bot Neerl 31: 441-447
53. DICKINSON DB, JE HOPPER, MD DAVIES 1973 A study of pollen enzymes involved in sugar nucleotide formation. In F. Loewus, ed, Biogenesis of Plant Cell Wall Polysaccharides, Academic, New York pp 29-48
54. LOEWUS F, M-S CHEN, MW LOEWUS 1973 The myo-inositol oxidation pathway to cell wall polysaccharides. In F. Loewus, ed, Biogenesis of Plant Cell Wall Polysaccharides, Academic, New York pp 1-27
55. LOEWUS F, C LABARCA 1973 Pistil secretion product and pollen tube wall formation. In F. Loewus, ed, Biogenesis of Plant Cell Wall Polysaccharides, Academic, New York pp 175-193
56. ROSENFIELD C-L, FA LOEWUS 1975 Carbohydrate interconversions in pollen-pistil interactions of the lily. In

DL Mulcahy, ed, Gamete Competition in Plants and Animals, North-Holland, Amsterdam pp 151-160

57. MORRISON IN, J KUC, TP O'BRIEN 1975 Histochemistry and fine structure of developing wheat aleurone cells. Planta 123: 105-116

58. McNEIL M, P ALBERSHEIM, L TAIZ, RL JONES 1975 The structure of plant cell walls. VII Barley aleurone cells. Plant Physiol 55: 64-68

59. SASAKI K, FA LOEWUS 1982 Redistribution of tritium during germination of grain harvested from _myo_-[2-^{3}H]-inositol- and _scyllo_-[R-^{3}H]inositol-labeled wheat. Plant Physiol 69: 220-225

60. MAITI IB, FA LOEWUS 1978 _myo_-Inositol metabolism in germinating wheat. Planta 142: 55-60

Chapter Ten

THE ABILITY OF WHEAT ALEURONE TISSUE TO PARTICIPATE IN ENDOSPERM MOBILIZATION

MICHAEL BLACK, JOHN CHAPMAN AND HELEN NORMAN

Department of Biology
Queen Elizabeth College
University of London
Campden Hill Road
London, W8 U.K.

INTRODUCTION

A distinctive property of the seed is that throughout its life a major part of metabolism is concerned, in one way or another, with the food reserves. During development of the seed, a high proportion of its increasing mass is storage material laid down by the activity of the multitude of enzymes which convert sucrose, amides and amino acids into starch, protein and triacylglycerols, in various proportions according to the species. This synthesis of reserves which dominates metabolism during development is not found again during the life of the seed, at least under normal circumstances. It is replaced, in the period following germination, by the transformation of reserves into sucrose, amino acids and amides, which move into the axis to support its growth, i.e. replaced by processes which in overall terms are the exact reverse of what happens in development. The seed thus displays a sharp temporal separation of metabolic processes involving these reserves--there is synthesis, and little or no breakdown as the seed develops; and mobilization, with virtually no synthesis, after the seed has germinated.

Important questions which can be raised are these: How is this separation in metabolism regulated? How are the activities of the mobilizing and synthesizing enzymes restricted to the two phases in the life of the seed? These questions are, of course, of the kinds that can be posed about the development of all organisms in which different patterns of metabolism are exhibited at different times, and though unequivocal answers are still not available, generalizations may be made in terms of regulation of gene activity etc. With regard to seeds, the questions are made rather more difficult to answer by our lack of knowledge about the control processes involved in food mobilization.

Hormonal effects have been invoked to explain the onset of mobilization in some dicotyledons but good evidence in favour of this interpretation is lacking and other possibilities, such as source/sink relationships, are perhaps more plausible.[1-3] The picture in seeds of some monocotyledons is clearer: here, the mobilizing enzymes such as α-amylase, proteases, 1,3-β-glucanase and ribonucleases are to a large extent (though not exclusively) derived from the aleurone layer which is induced to synthesize and secrete them by a regulatory factor, gibberellin, emanating from the embryo. The action of this hormone has been well documented[1] and it is now clear that aleurone cells of many strains and cultivars of grasses such as rice, barley, wheat and wild oats are almost completely incapable of producing the enzymes needed for mobilization of the endosperm reserves unless they are exposed to gibberellin. We have therefore chosen to approach the question of how the activity of mobilizing enzymes is confined to the post-germination phase by examining the control of aleurone layer activity in wheat, *Triticum aestivum* L.

RESPONSIVITY OF THE WHEAT ALEURONE LAYER TO GIBBERELLIN

The control system operating in the aleurone layer offers one possible means by which the mobilizing enzymes might be temporally confined, i.e. by restricting the regulatory factor, gibberellin, to the post-germination phase of the seed's life. It happens, however, that during their development, wheat grains[4-6] (and barley[7]) contain substantial amounts of the regulator. Usually, two separate peaks of extractable gibberellin occur at various times after anthesis, depending on the cultivar and growing conditions (Fig. 1). In 24-day old grains, 90% of the gibberellin is in the endosperm[6] and corresponds to a concentration of approximately 10^{-7}mol dm^{-3}, so if we assume an equivalent distribution between the aleurone layer and the starchy endosperm cells, the aleurone cells in the developing grains are exposed to a dosage of gibberellin which is close to saturation value, at least for mature wheat. If these cells responded to the hormone copious quantities of α-amylase, protease and other hydrolases would be produced, an event which obviously could drastically interfere with the laying down of the various reserves which is proceeding at this time.

We find, however, that for a substantial period during the development of the wheat and barley grain the aleurone tissue is completely insensitive to gibberellic acid, as indicated by the failure of de-embryonated grains to produce α-amylase when exposed to the growth regulator[8-13] (Fig. 2). Only when the grain begins to ripen does the aleurone tissue acquire sensitivity towards applied gibberellin. By this time, however, the endogenous gibberellin is barely detectable,[6] but even if the regulator were present at this stage

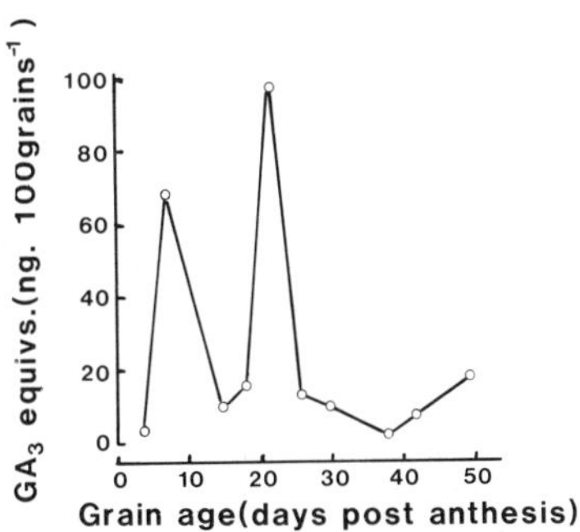

Fig. 1. Gibberellin content of developing wheat grains (cv. Champlein). Gibberellin-like substances assayed by the barley aleurone test. From Mitchell.[6]

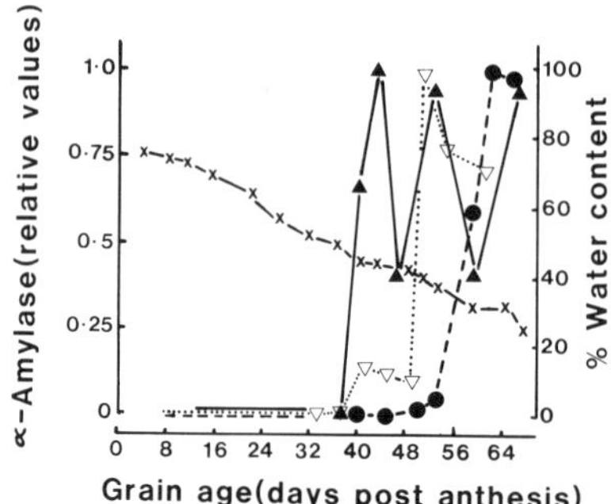

Fig. 2. Sensitivity to gibberellic acid of aleurone tissue of developing wheat grains. Throughout their development samples of grains were collected, de-embryonated and tested for their response to 10^{-5}mol dm^{-3} gibberellic acid. Symbols: —▲— ww15 (from King[9]), ---▽---cv. Champlein (from Armstrong[14]),---●--- cv. Sappo (from Armstrong et al[12]). Individual points for ages below 34 days are ommitted. The curve for grain water content in cv. Sappo is also shown (x — x — x). No detectable α-amylase was produced in the absence of GA_3.

the aleurone cells could not respond to it because they are fairly dehydrated! It seems, therefore, that throughout its development the storage tissue is protected against the possibility of attack by aleurone-derived hydrolases because the aleurone cells simply are unable to respond to the regulator. Thus, temporal restriction in the production of mobilizing enzymes by the aleurone layer is achieved by variation in the sensitivity of this tissue to the regulatory factor, gibberellin.

INDUCTION OF SENSITIVITY TO GIBBERELLIN

The question now arises as to how the aleurone cells acquire their sensitivity to gibberellin. It is possible that onset of sensitivity may simply be part of a developmental program reached when cells achieve a certain age--apparently coincident with harvest ripeness of the grains. But alternatively, it may be that ability to respond to the growth substance results from some other critical change in the grain which is not specifically age dependent. As we saw from Fig. 2 sensitivity of the aleurone cells sets in at the end of development when the grains have become relatively dry. That drying and not grain age is the experience which sensitizes the aleurone tissue to gibberellin is shown by the fact that in very young grains (e.g. 20 days after anthesis) this tissue can be converted into a hormone-sensitive condition by dehydration prior to exposure to gibberellin[8-12] (Fig. 3). We should note that to demonstrate this point grains must be removed from the ear, and it has been suggested that this detachment from the mother plant may have some influence, perhaps by curtailing metabolic processes in the grain which depend upon the continuous influx of carbohydrate.[10] Detachment alone cannot, however, account for the sensitization of the aleurone layer because this does not occur in isolated grains held in a liquid medium for several days.[12] We conclude, therefore, that loss of water from the aleurone tissue is the prerequisite for the development of sensitivity to gibberellin. Drying is thus the critical experience which confers upon the aleurone cells the ability to respond to gibberellin and produce α-amylase, and hence it converts the incompetent tissue into one which can participate in the mobilization of food reserves. Dehydration, and not some inbuilt developmental program, is used by the grain to recognize that it has reached the stage in

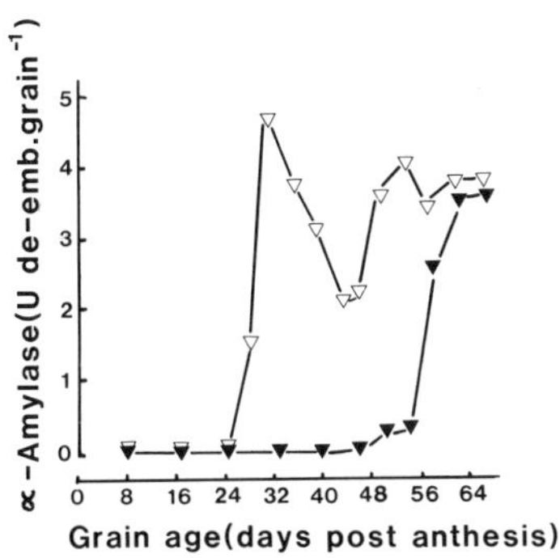

Fig. 3. Sensitization of wheat aleurone tissue to gibberellin by drying. Grains (cv. Sappo) of different ages were dried for 2 to 3 days at room temperature to approximately 15% water content, embryos were removed, and the de-embryonated grains were incubated in 2×10^{-6} mol dm^{-3} GA_3 for 96 h, after which time the total α-amylase activity was determined (no α-amylase in the absence of GA_3). Symbols: —▼— non-dried grains, and —▽— dried grains. After Armstrong et al.[12]

its life when utilization of its food reserves need no longer be blocked. As soon as gibberellin appears--an event which follows reimbibition and germination--production of α-amylase can commence and starch breakdown ensues.

Mechanism of Action of Drying

Acquisition of sensitivity to gibberellin by the aleurone tissue occurs fairly abruptly as the grain ripens. In the examples shown in Fig. 2 it takes place over a period of 3 to 5 days, during a time when the water content of the grain as a whole is still falling, and has dropped, in cv. Sappo, for example to a value of 35 to 30% of the fresh weight. It seems possible, therefore, that sensitization to gibberellin does not develop gradually as the cells dry but rather that some critical level of drying must be reached. This point is confirmed directly by controlling the drying of grains down to different water contents, followed by introduction of the de-embryonated grain into buffer containing gibberellic acid (Fig. 4). It is clear that drying to below 30% water content must occur before any sensitivity to gibberellin is acquired. The critical water level of 25 to 30%, which has been confirmed for two cultivars of wheat, Sappo and Champlein,[12,14] applies of course to the grain as a whole. It is possible that small differences could occur among different parts and

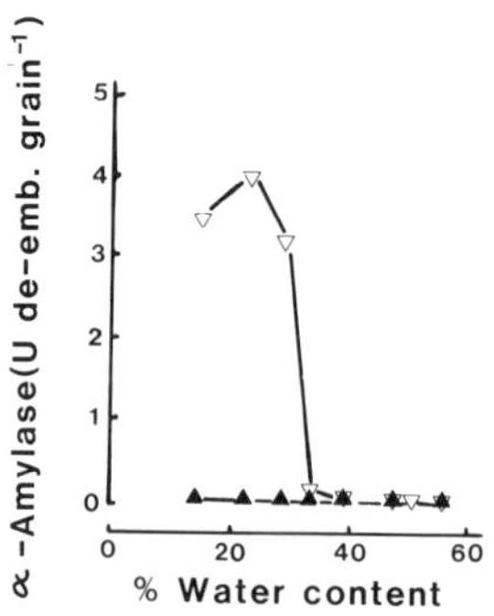

Fig. 4. Water content and sensitization to gibberellin. Thirty-two-day old grains (cv. Sappo) were dried to different water contents and afterwards de-embryonated. The total α-amylase made in response to 10^{-5}mol dm^{-3} GA_3 over 96 h was measured, —▽— with GA_3; —▲— without GA_3. From Armstrong et al.[12]

that the aleurone layer itself may be somewhat more or less dehydrated.

The necessity for dehydration to a critical content of water might give us a clue as to the cellular changes which precede sensitization and which, presumably, form the basis of the mechanism of action of drying. The value of about 25% water is that at which cell membranes are thought to undergo structural changes. These might be the severe disruption of the bilayer as the phospholipid molecules assume the so-called hexagonal array around channels of water (a change known to occur in artificial phospholipid membranes),[15] or the thinning and disorganization which takes place in some tissues, such as seeds of _Glycine max_.[16,17] Either change would obviously alter both the environment and the relative positions of membrane components such as proteins, and possibly bring about conformational changes in them. Could alterations in cell membranes therefore be responsible for the transformation of the aleurone cell from a GA-insensitive to a GA-sensitive state?

TEMPERATURE-CONTROLLED SENSITIZATION OF ALEURONE CELLS

The structure of cell membranes can be altered by temperature as well as by drying. Characteristically, membranes undergo transitions if certain critical temperatures are exceeded, when ordering of the phospholipids switches from a gel to a liquid crystalline state.[18] Because membranes are heterogeneous in their lipid content and because they also contain protein, highly localized changes may occur at certain physiological temperatures so that the membrane as a whole would contain both gel and liquid crystalline states simultaneously. When the transitions happen, changes take place in the permeability properties of the membrane, in the diffusion of lipids and protein within the membrane and in the orientation and conformaion of proteinaceous components. There is good evidence that the physiology of seeds can be affected by temperature-induced transitions in the state of cell membranes. One striking effect, documented by the work of Hendricks and Taylorson,[19] concerns the breaking of dormancy, in particular the capacity of seeds to respond to low levels of the form of phytochrome which promotes germination.

The concept that structural changes in cell membranes form the basis of the sensitization mechanism in the aleurone cells gains strong support from the effect of certain temperatures on this tissue. Exposure of undried, immature, de-embryonated wheat grains to temperatures above a critical value sensitizes the aleurone tissue to gibberellic acid so that it is capable of producing α-amylase when subsequently exposed to the growth regulator.[13] Effective temperatures lie above 27°C, a temperature of 26°C being completely ineffectual (Fig. 5). In intact cells, the magnitude of the sensitization increases up to 30°C, falls (for reasons which are not understood) and rises again up to 35°C. Sensitization of protoplasts isolated from undried wheat aleurone cells is also achieved by the same temperatures and such 'transformed' protoplasts respond to GA_3, as far as α-amylase production is concerned, almost as well as do dried or temperature-treated intact cells (Fig. 5). Thus, populations

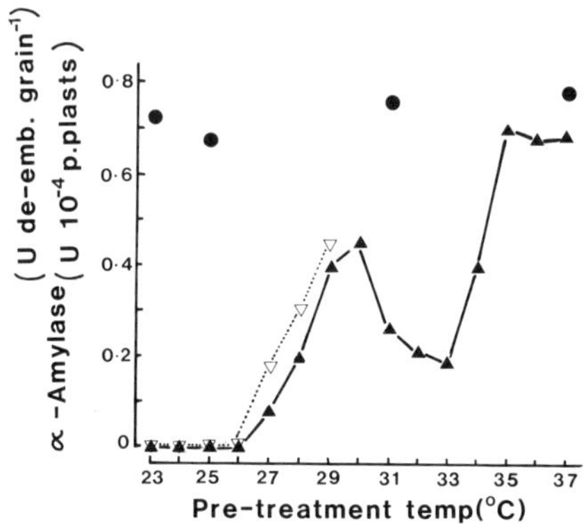

Fig. 5. Induction of sensitivity to gibberellin by different temperatures. De-embryonated grains (21-day old, non-dried cv. Sappo) were incubated for 12 h at different temperatures in the absence of GA_3. They were then transferred to 2×10^{-6}mol dm^{-3} for 48 h at 25°C, after which the total α-amylase was measured. As controls, dried grains were given the same temperature treatments prior to incubation in GA_3. Isolated protoplasts from non-dried grains were temperature treated for 12 h in the absence of GA_3 and transferred to 10^{-8}mol dm^{-3} GA_3 for 72 h at 25°C when the total α-amylase was determined. Symbols: —▲— non-dried grain, ● dried grain, and —▽— isolated protoplasts. No α-amylase was detectable in the absence of GA_3. From Norman et al[13] and unpublished results of the authors.

of aleurone cells and isolated protoplasts are obtainable whose sensitivity to gibberellin can be manipulated, simply by a treatment with temperatures above 27°C.

The occurrence of a critical temperature for sensitization prompts the interpretation that a membrane transition is taking place. This interpretation is further supported by two observations: (a) Leakage of amino acids from undried aleurone cells is enhanced by temperatures above the critical 27°C,[13] indicating that permeability of the plasma membrane is affected, (b) The negative electrical charge of the cell surface (zeta potential) of protoplasts isolated from undried aleurone cells abruptly increases when the same critical temperature is exceeded (Norman, Black and Chapman, unpublished studies). This provides direct evidence that a structural alteration occurs in the plasma membrane which reveals negatively charged groups, possibly of phospholipid and/or protein. It seems clear, then, that changes in cell membranes, perhaps in the plasma membrane alone but certainly including it, are responsible for the profound alteration in the capacity of the aleurone layer to respond to gibberellic acid which is brought about by certain temperatures.

Time Requirement at Above the Critical Temperature

The transition from the gel to the liquid crystalline state of a membrane occurs virtually instantaneously when the critical temperature is exceeded. It is clear that this happens in the present case since the change in zeta potential of the protoplast surface is exhibited as soon as the cell is exposed to a temperature above 27°C. It is important to note, however, that the charge change is fully reversible for about 5 h, and the electronegativity reverts to the 'base-line' value if the temperature falls to 25°C during this time. But after 5 h at above 27°C the new zeta potential becomes fixed and is unaffected by a fall in temperature (Fig. 6).

It now becomes pertinent to ask the following question: if the membrane transition and the change in cell surface charge occur as soon as the critical temperature is exceeded, does sensitization to GA_3 (insofar as α-amylase production is concerned) take place equally rapidly? Unfortunately, this is difficult if not impossible to determine since enzyme production is a fairly long-term process which, even if initiated quickly, is not apparent until after many hours have passed. However, the other aspect of the kinetics of plasma

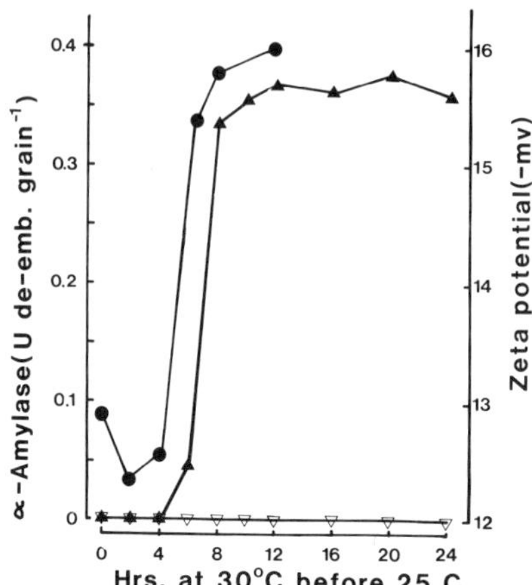

Fig. 6. Time requirements for the 'high-temperature' effect. De-embryonated grains (21-day old, cv. Sappo) were held in the absence of GA_3 for periods of time at 30°C, and afterwards incubated for 48 h in 10^{-6}mol dm^{-3} GA_3 at 25°C when the total α-amylase was determined. Isolated protoplasts were held at 30°C for periods of time and the zeta potential measured after transfer to 25°C. GA_3 was absent from the protoplast treatments. Symbols: —●— protoplast zeta potential, —▲— α-amylase (with GA_3), —▽— α-amylase (without GA_3). From Norman et al[13] and unpublished results of the authors.

membrane change--the time required for its stabilization--is approachable in terms of sensitivity. To become permanently sensitized to GA_3, aleurone cells must experience more than 5 h at temperatures above 27°C. In the curve shown in Fig. 6 some transformation is apparent in cells which have received 30°C for 6 h but the maximum effect is achieved by 8 h of treatment. This close agreement between the length of time above 27°C needed for stabilization of the increased electronegativity of the plasma membrane and that required for sensitization to GA_3 is evidence that the alteration in plasma membrane charge is intimately connected with the acquisition of sensitivity to the growth regulator.

Biochemical Changes Occurring at the Higher Temperatures

What happens during the 6 or more hours at above 27°C which stabilizes the altered surface charge and is required for inducing sensitivity to GA_3? Certain long-term events must occur in response to the increased membrane fluidity which clearly affect the surface membrane itself, since afterwards its electronegativity remains permanently increased. Adjustments to the composition of membranes are known to occur in plants, animals and microorganisms in response to temperature shifts.[20,21] These adaptive responses (homeoviscous adaptation) largely consist of qualitative and quantitative changes in the phospholipids, particularly in the degree of saturation of the fatty acids. Such compositional changes do, indeed, take place in both intact aleurone cells and isolated aleurone protoplasts in response to the supra-critical temperature, there being an increase in the proportion of saturated to unsaturated fatty acids, particularly in phosphatidyl choline (Fig. 7). Moreover, substantial alteration in phospholipid composition is detectable only after 6 h at 30°C, which is the time needed for the cells to become permanently sensitized to GA_3 and to fix the altered zeta potential. It seems clear then, that an important change taking place in the cell membranes, including the plasma membrane, during the temperature sensitization process is the modification of phospholipids to make them more saturated. This presumably stabilizes both the increased electronegativity of the cell surface and the induced sensitivity to GA_3.

Interestingly, similar changes in phospholipid composition occur in response to drying. When immature aleurone cells are dried and then rehydrated, the proportion of saturated to unsaturated fatty acids in the membrane phospholipids increases during the first few hours of re-imbibition. This also happen when dry mature grains take up water (Norman, Black and Chapman, unpublished studies). Thus, the two experiences which cause aleurone tissue to become sensitive to gibberellic acid--drying and a critical temperature--both result in closely similar alterations in the cells' membrane phospholipids.

Events Following the High-temperature Effects

The standard practice in assessing the effects of supra-critical temperatures is to follow any high temperature treat-

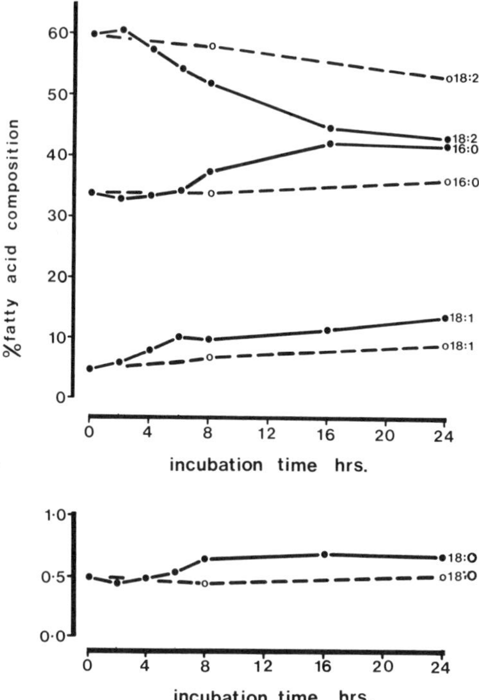

Fig. 7. Changes in fatty acid content in response to supra-critical temperature. De-embryonated grains (24-day old, cv. Sappo) were incubated at 25° and 30°C (minus GA_3). At different times, aleurone layers were removed, phospholipids extracted and separated by thin-layer chromatography. Phosphatidylcholine, the fatty acid content of which is shown here, was removed and analyzed by gas-liquid chromatography. Symbols: —○— incubation at 25°C, and —●— incubation at 30°C.

ment with the incubation of the de-embryonated grains or isolated protoplasts in GA_3 at 25°C. However, if the temperature remains at above 27°C throughout the whole treatment, i.e. including the incubation in GA_3, the aleurone cells are unresponsive to the growth regulator--the sensitization treatment apparently fails. This is because the supra-critical temperature must be followed by a period at a lower temperature (<approximately 28°C) for the sensitization mechanism to succeed. The time requirements for the high

and low temperature treatments are, however, of completely different orders of magnitude. Several hours at above 27°C are needed, as we have already seen, but after this a few seconds below 28°C suffice to complete the sensitization process (Table 1). Events occurring over this exceedingly short duration are unlikely to be metabolic, of the sort requiring synthesis or insertion of novel compounds into the membrane. Instead, they are probably rapid configurational or conformational changes in membrane components. These events finally fix the membrane in a GA-sensitive state, a condition which is retained at least over several days prior to introduction of the growth regulator.

Table 1. Need for interpolated low temperature for sensitization of aleurone cells to gibberellic acid.

TREATMENT SEQUENCE				
12 Hour pre-treatment temperature ($-GA_3$)	Time in ice ($-GA_3$)	Time at 25°C ($-GA_3$)	48 Hour incubation temperature ($+GA_3$)	α-Amylase activity
°C	sec	min	°C	U $grain^{-1}$
25	0	0	25	0.000
30	0	0	25	0.358
30	0	0	30	0.000
30	8	0	30	0.000
30	14	0	30	0.255
30	14	1	30	0.451
30	14	60	30	0.471

Grain temp. 8 sec in ice = 27°C
14 sec in ice = 25°C

De-embryonated grains were given different temperature treatments before incubation for 48 h in 10^{-6}mol dm^{-3} GA_3. Grain temperature, monitored by catheter probes, dropped to 27°C after 8 sec in ice, and to 25°C after 14 sec in ice. Note that grains held at 25°C and 30°C throughout fail to respond to GA_3, but a low-temperature interruption facilitates the response to 30°C.

From Norman et al.[13]

A MODEL FOR MEMBRANE TRANSFORMATION AND THE IMPLICATIONS FOR GA_3 ACTION

The foregoing observations contribute a substantial body of evidence suggesting that aleurone cells are transformed from a GA_3-insensitive to a GA_3-sensitive state by conversion of cell membranes, particularly the plasma membrane. All of this evidence, summarized in Table 2, will be incorporated in the following hypothesis which attempts to explain the changes occurring in the membranes and how these endow the cell with competence to respond to gibberellin.

At physiological temperatures, much of the biomembrane is probably in the liquid crystalline state except in certain domains around protein molecules.[22] It is in these regions that transitions most likely take place as soon as the critical 27°C is exceeded. The first measurable effect of this is an immediate change in zeta potential--the increase in electronegativity of the cell surface. The basal negative

Table 2. Evidence for membrane changes during sensitization of aleurone tissue to gibberellin.

Physiology
α-Amylase
Drying-rehydration (<30% water)
Critical temperature
Heat ≥ 27°C - slow response
Cool ≤ 28°C - rapid and irreversible
Amino leakage (≤ 27°C)
Composition and structure
Phospholipids
Increase in saturation ≥ 27°C and after dehydration.
Increase in phosphatidylcholine, phosphatidylinositol.
Protoplast surface charge
More electronegative ≥ 27°C (fast response)
Electronegativity 'fixed' after 6 h ≥ 27°C
Membrane proteins change ≥ 27°C
Gibberellin action
Electronegativity of sensitized protoplasts reduced.

charge of the plasma membrane is presumably due to phospholipid head groups and to proteins, both of which may contribute to the altered charge when the membrane is disturbed, though the effects of these two constituents may not be of equal magnitude; the exposure of negatively charged groups on proteins, for example, might account for the major part of the enhanced electronegativity (stage A, Fig. 8). A consequence of the altered ordering of the plasma membrane components is its increased permeability as revealed by the leakage of amino acids from the cells. Another expected result would be that diffusion of proteins could occur more easily within the 'liquid' phospholipids and possibly approach each other. Initially, no biochemical adjustments to the membrane follows these events and if the temperature is reduced to say, 25°C, the zeta potential immediately reverts to the original value as the membrane components assume their original positions. When the increased fluidity is retained for several hours, however, the cell responds by homeoviscous adaptation, altering the fatty acid composition of the cell membranes to adjust their viscosity. After about 6 h, this adaptation is detectable in the increased level of saturation of the phospholipid fatty acids, an effect which increases with time spent above 27°C (stage B, Fig. 7). Thus, new phospholipid composition stabilizes the zeta potential, presumably by fixing negatively charged groups (e.g. of proteins) in an exposed position. But this is still insufficient to sensitize the cell to gibberellic acid. The final event which consolidates the sensitization process is the transient experience of a temperature less than approximately 28°C. This probably causes the new saturated phospholipids to assume the more ordered, gel state, a physical change which has a profound effect. One might speculate that this stage could cause the association of two protein sub-units and thus generate a receptor protein or 'composite' for GA_3 (stage C, Fig. 8). Preliminary evidence obtained from the iodination of surface proteins with ^{125}I does, indeed, point to the greater exposure of certain membrane proteins after the sensitization process has been completed; and it seems also that novel proteins arise in the membrane (Norman, Black and Chapman, unpublished studies).

We postulate these events as possible ones which transform the aleurone cell into a GA_3-sensitive condition. In all the experimental work discussed above sensitivity to GA_3 has been assessed by the eventual production and release of α-amylase. However, the response of isolated protoplasts to

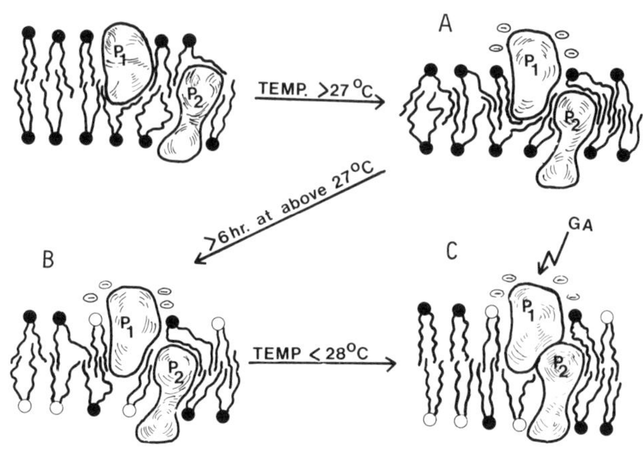

Fig. 8. Postulated changes in the plasma membrane involved in sensitization of the aleurone cells to gibberellin. In A, phospholipids immediately become more fluid at temperatures above the critical 27°C. The resultant thinning of the membrane causes the protein (P_1) to protrude, revealing negative changes on the outer surface of the plasma membrane. Protein P_2 moves closer to P_1. After 6 h, homeoviscous adaptation begins to change the phospholipids (B). With a lowering of the temperature to below 28°C, the phospholipids rapidly become more ordered (viscous) causing the irreversible association of P_1 and P_2 which now form a complex. This, when interacting with GA_3, subsequently induces responses in the aleurone cells.

GA_3 provides other, direct evidence that the cell surface of transformed cells only can interact with GA_3. When the growth regulator is added to transformed protoplasts, there is an immediate drop in zeta potential: gibberellic acid has no such effect on non-sensitized cells (Norman, Black and Chapman, unpublished studies). Thus, the primary receptor site for GA_3 is the plasma membrane, as is indicated in the model (stage D, Fig. 8). Whether this site controls the transcription processes taking place during α-amylase synthesis or regulates secretory activity is not yet clear.

This model, admittedly speculative but fully consistent with the experimental findings, can explain how aleurone tissue is transformed by temperature treatments into a GA_3-sensitive state. In nature however, the transformation is accomplished by dehydration. One can envisage that disruption of the membrane during desiccation might allow membrane proteins to associate to form a receptor, or even permit the entry of novel proteins into the membrane during the early stages of rehydration. Hence, events depicted under stage C, Fig. 8 could possibly take place: presumably these, too, are fixed by the new phospholipid composition which follows the rehydration of the aleurone cells.

CONCLUSIONS

These studies show that the aleurone tissue of wheat is normally unable to participate in the mobilization of reserve starch until the grain is fully mature. Maturity is recognized by the loss of water which accompanies ripening but it can be simulated by premature dehydration or by certain temperature treatments. Both of these factors cause profound changes to occur in the aleurone cells, central to which is the plasma membrane and its conversion into a GA_3-receptive state. We have used the response to temperature as a means of probing into the membrane changes which are involved in the transformation of the aleurone layer into a tissue which can produce α-amylase in response to gibberellin. So far, we have concentrated on this one enzyme, but it is important to determine if the production of proteases, ribonucleases, and pentosanases falls under the same type of regulatory control, and similarly depends upon the completion of dehydration.

Questions obviously can be raised about the possible application of these findings to other systems. In all seeds there is a temporal control of mobilizing activity, to which we have already referred; and one would like to know if in these cases, too, the experience of dehydration is used as a means of recognizing that a stage has been reached when mobilization is permitted. Moreover, one may ask if the nature of the plasma membrane exerts ultimate control over the capacity of a seed to switch from reserve synthesis to reserve mobilization, as it does in the case of the wheat grain aleurone cells.

REFERENCES

1. BEWLEY JD, M BLACK 1978 Physiology and Biochemistry of Seeds. Springer-Verlag, Berlin, Heidelberg, New York pp 306

2. DAVIES H, P SLACK 1981 The control of food mobilization in seeds of dicotyledonous plants. New Phytol 88: 41-51
3. DAVIES H, J CHAPMAN 1979 The control of food mobilization in seeds of *Cucumis sativus* L. II. The role of the embryonic axis. Planta 146: 585-590
4. WHEELER AW 1972 Changes in growth substance contents during growth of wheat grains. Ann Appl Biol 72: 327-334
5. GASKIN P, PS KIRKWOOD, JR LENTON, J MACMILLAN, ME RADLEY 1980 Identification of gibberellins in developing wheat grain. Agric Biol Chem 44: 1589-1593
6. MITCHELL BA 1980 The control of germination ability in developing wheat grains. PhD Thesis, University of London
7. MOUNLA MA KH, G MICHAEL 1973 Gibberellin-like substances in developing barley grain and their relation to dry weight changes. Physiol Plant 29: 274-276
8. EVANS M, M BLACK, J CHAPMAN 1975 Induction of hormone sensitivity by dehydration is one positive role for drying in cereal seed. Nature (London) 258: 144-145
9. KING RW 1976 Abscisic acid in developing wheat grains and its relationship to grain growth and maturation. Planta 132: 43-51
10. NICHOLLS PB 1979 Induction of sensitivity to gibberellic acid in developing wheat caryopses: effect of rate of desiccation. Aust J Plant Physiol 6: 229-240
11. KING RW, M GALE 1979 Pre-harvest assessment of potential α-amylase production. Cereal Res Commun 8: 157-165
12. ARMSTRONG C, M BLACK, J CHAPMAN, HA NORMAN, R ANGOLD 1982 The induction of sensitivity to gibberellin in aleurone tissue of developing wheat grains. I. The effect of dehydration. Planta 154: 573-577
13. NORMAN H, M BLACK, J CHAPMAN 1982 The induction of sensitivity to gibberellin in aleurone tissue of developing wheat grains. II. Evidence for temperature-dependent membrane transitions. Planta 154: 578-586
14. ARMSTRONG C 1980 Gibberellin action in the aleurone layer of developing wheat. PhD Thesis, University of London
15. SIMON EW 1974 Phospholipids and plant membrane permeability. New Phytol 73: 377-420
16. SEEWALDT V, DA PRIESTLY, AC LEOPOLD, GW FEIGENSEN, F GOODSAID-ZALDUONDO 1981 Membrane organization in soybean seeds during hydration. Planta 152: 19-23

17. CHABOT JF, AC LEOPOLD 1982 Ultrastructural changes of membranes with hydration in soybean seeds. Amer J Bot 69: 623-633
18. MELCHIOR DL, JM STEIM 1976 Thermotropic transitions in biomembranes. Annu Rev Biophys Bioeng 5: 205-238
19. HENDRICKS SB, RB TAYLORSON 1979 Dependence of thermal responses of seed on membrane transitions. Proc Natl Acad Sci USA 76: 778-781
20. LYONS JM 1973 Chilling injury in plants. Annu Rev Plant Physiol 24: 445-466
21. SINENSKY M 1974 Homeoviscous adaptation: a homeostatic process that regulates the viscosity of membrane lipids in _Escherichia coli_. Proc Natl Acad Sci USA 71: 522-525
22. OLDFIELD E, D CHAPMAN 1972 Dynamics of lipids in membranes: heterogeneity and the role of cholesterol. FEBS Lett 23: 285-297

Chapter Eleven

MOBILIZATION OF SEED INDOLE-3-ACETIC ACID RESERVES DURING GERMINATION

ROBERT S. BANDURSKI

Department of Botany and Plant Pathology
Michigan State University
East Lansing, Michigan 48824-1312

INTRODUCTION

Seeds contain conjugates of indole-3-acetic acid (IAA),[1] the gibberellins, the cytokinins, and abscisic acid. Hormones other than IAA will not be reviewed in this work and have been reviewed elsewhere (e.g. Ref. 2). In addition, seeds contain auxins other than IAA as, for example, phenylacetic acid[2] and the 4-chloroindole-3-acetic acid of Gander and Nitsch et al, Marumo et al, and Engvild et al, and have been previously reviewed[2-4] and will not be considered here. Further, the bulk of this work will be addressed to the seedling of *Zea mays* since this has been the major subject of our research. It is hoped that by means of these omissions that greater attention and concentration may be centered on the many mysteries and uncertainties concerning the seed as a source of IAA for the growing vegetative tissue of the shoot. The greatest mysteries, of course, are those that have frequently been mentioned throughout this volume including how the shoot "tells" the seed at what rate the reserves should be mobilized and transported, and in the case of the monocotyledonous seedling how materials are moved from the endosperm into the scutellum, from the scutellum into the vascular stele, and from the stele into the vegetative tissue. We can only allude to these important determinants of the early stages of shoot growth in the hope they receive further attention. In the spirit of this belief I hope you will tolerate an introduction that is almost as long as the body of the text.

Knowledge that the seed could serve as a source of auxin for the shoot dates back, at least, to Cholodny.[5] Cholodny excised small blocks of endosperm tissue, moistened the blocks, and then applied the moistened block to young grass seedlings. Curvatures developed owing to the more rapid growth on the side to which the block had been applied, and this curvature served as a semi-quantitative assay for growth hormone in the block. Further, Cholodny found that moistening the block with ethanol, rather than water, decreased, or even eliminated the curvature. Thus, Cholodny is the discoverer of the seed auxin precursor and, although he did not realize it, the discoverer of enzymes which hydrolyzed the seed auxin precursor to make it into an active auxin.

Then there developed a large and elegant number of studies concerning the concentration and chemical properties of the seed auxins. The chemical properties studied included the solubilities of the seed auxin precursors and the conditions required to hydrolyze the seed reserves. Almost no attempts were made to chemically characterize these compounds and this appears to be owing to three findings: First, the demonstration by Berger and Avery[6] that the auxin complex was a high molecular weight compound--probably a protein. The purification methods then available were clearly inadequate to chemically characterize a high molecular weight compound. Second, was the knowledge provided by Kögl et al,[7] that IAA was an auxin. Even though it was only believed to be a "heteroauxin", IAA was more attractive to experimentalists than was some poorly defined protein. Lastly, was the discovery by Thimann[8] that tryptophan could be converted to IAA. This well-defined conversion proved more attractive than studies on a high molecular weight protein whose auxin yield, in any event, was probably owing to conversion of tryptophan in the protein to IAA. This was supported by Wildman et al,[9] and by Schocken,[10] and had deadly consequences for any further study of IAA conjugates. Thus, until very recently, only nominal efforts have been directed towards understanding the utilization of seed auxin reserves other than tryptophan during germination.

CHEMICAL COMPOSITION OF SEED IAA RESERVES

The chemical composition of the IAA conjugates of *Zea mays* has been reviewed elsewhere but is summarized in Table 1. As can be seen, one-half of the conjugates are low molecular weight compounds and only one-half are the high molecular weight compounds studied by Berger and Avery. It was Labarca et al [11] who made the finding that one-half of the conjugates are low molecular weight, a finding which led to elucidation of the structures of the compounds.

Knowing the structures of the *Zea* conjugates, there still remains several important questions. First, why should *myo*-inositol be the alcohol component of the IAA ester? Loewus and Dickinson[12] have addressed the subject of *myo*-inositol metabolism and Agranoff and Bleasdale[13] have focused attention on - "Why *myo*-inositol?" Although *myo*-inositol is abundant there are other more abundant carbohydrates. Unpublished studies of A. Ehmann showed that the same family of IAA-*myo*-inositol and IAA-*myo*-inositol glycosides occurs in sweet and field corn, in pop corn, and in *Teosinte* and *Trypsicum*. Work of P. J. Hall[14] has shown that IAA-*myo*-inositol also occurs in rice (*Oryza sativa*). Whether *myo*-inositol esters of IAA

Table 1. A summary of the amounts of indole-3-acetic acid and its conjugates in *Zea mays* kernels.

Compound	Amount in dry seed mg/kg	Percent of total
Indole-3-acetic acid	0.5	0.8
Indole-3-acetyl *myo*-inositols	10	15
Indole-3-acetyl *myo*-inositol arabinosides	15	23
Indole-3-acetyl *myo*-inositol galactosides	5	8
{ Di and tri (indole-3-acetyl) *myo*-inositols; 2-0, 4-0 and 6-0 IAA glucose }	0.2	0.3
Indole-3-acetyl glucan	35	53

occur in other cereals, or even in dicotyledonous plants, is simply unknown.

The high molecular weight IAA conjugate, thought by Berger and Avery to be a protein, has been studied by Piskornik[15] and shown to be an IAA-glucan. The linkage of the glucosyl residues is β(1→4) and the chain length, assuming one IAA per glucan, is between 8 to 50 glucosyl residues. The glucan, as well as all the IAA conjugates, has been shown to be located in the endosperm by Piskornik[16] and shown by Ueda[17] to "disappear" during germination.

Only two other seeds have been studied with regard to chemical characterization of the IAA conjugates including _Avena sativa_ studied by Percival et al[17] and _Glycine max_ studied by Cohen.[19] The _Avena_ conjugate (see Table 2) proved to be mainly a high molecular weight IAA-glycoprotein. The glycan was a lichenin glucan with mixed (1→3),(1→4)-β-linkages and this, of course, is the familiar high viscosity glucan of oatmeal and porridge. The protein moiety had a pK_a of 4.2 or 4.7 so that two distinct IAA-gluco-proteins could be isolated. Both species became highly labile after a certain degree of purification so that the IAA was easily destroyed and thus a more detailed characterization could not be accomplished. There were, in addition, small amounts of a low molecular weight IAA-auxin closely resembling IAA-aspartate. This is of singular interest because the IAA conjugates of the vegetative shoots of _Avena_ are mainly amide linked.[20]

A major conjugate of soybean has been shown to be IAA-L-aspartate. By using ^{14}C-labeled IAA-L-aspartate as a tracer and as an internal standard, Cohen was able to demonstrate that a) IAA-L-aspartate occurs as an endogenous compound (previous demonstrations of the occurrence of IAA-aspartate were always made following external application of IAA), and b) it was an L-aspartate.[19] Although we were not aware of it at the time of our experiments, evidence had been provided by Zenk that it was an L-aspartic acid.[21] This is of particular interest since Hangarter et al[22] have shown that it is the L-amino acid conjugates which are metabolized.

A somewhat broader collection of plant species were surveyed _quantitatively_ for their content of IAA-conjugates.[20] In these cases no attempt was made to structurally characterize the conjugates except to determine whether the IAA was ester-linked or amide-linked. Two generalities emerge, the

Table 2. A summary of indole-3-acetic acid conjugates in cereal plants.[14]

Plant	High MW		Low MW	
	Nature	Quantity %	Nature	Quantity %
Avena sativa	IAA-glucoprotein[a]	80	Unknown	20
Oryza sativa	Unknown	90 (estimated)	IAA-*myo*-inositol	10
Zea mays	IAA-cellulosic glucan[b]	50	IAA-*myo*-inositol and IAA-*myo*-inositol glucosides	50

[a] IAA linked to a glucoprotein (1 → 3),(1→4)-β-linked D-glucose residues.
[b] IAA linked to a glucan containing (1 → 4)-β-linked D-glucose residues.

most important being that all seeds and vegetative tissues examined had most of their IAA in conjugated form, and, that the cereals contain primarily esters whereas the dicotyledonous plants contain mainly amide linked conjugates.

Still, the data are too few to permit many conclusions. For example, an important question is whether the IAA conjugates of the seed will reflect the IAA conjugates found in the shoot. The case of *Avena* would indicate that this is *not* the case since the seed contains a high molecular weight ester whereas the shoot contains an amide linked IAA. Our postulate is that the high molecular weight conjugates, such as occur in *Avena* and in *Zea*, will prove to be sources for IAA in the kernels and are needed for some germination purpose not yet understood. It is in the low molecular weight conjugates, that we may find that the composition of the seeds will reflect that of the vegetative shoot tissue, since the low molecular weight forms are more likely to be transport forms. What function IAA released by hydrolysis of a high molecular weight conjugate in the seed might have is totally unknown.

HYDROLYSIS OF IAA CONJUGATES IN THE ENDOSPERM

As mentioned above, Cholodny knew that moistened endosperm tissue could release a growth promoting substance and, in fact, Hatcher[23] and others (cf. Ref. 1) knew that the bound IAA of the kernel disappeared during germination.

Ueda and Bandurski[17] determined that the IAA conjugates of *Zea* kernels disappeared at a rate of about 1% per h during the first 96 h of germination. Both the IAA-*myo*-inositol esters and the high molecular weight IAA-glucan were lost at about the same rate except that the IAA-*myo*-inositol glycosides were lost more slowly. The amount of IAA ester being transported to the shoot is very small compared to the rate at which the conjugates are being lost and this will be discussed in the following section.

Epstein et al[24] were the first to demonstrate conclusively that conjugate loss during germination represented conjugate hydrolysis and that free IAA was a product. This was accomplished by applying ^{14}C-labeled IAA to a cut endosperm surface. Since the endosperm is soft in consistency--like Halavah--and since experiments with a dye indicated fairly uniform coloration of the endosperm, it is a reasonable first approximation that the ^{14}C-labeled IAA will diffuse fairly uniformly into the endosperm tissue. What Epstein et al observed was that the specific activity of the applied IAA decreased as a function of time. The rate of decreased specific activity exhibited first order kinetics and so it was possible to calculate that new IAA was being produced at a rate of about 100 $pmol \cdot kernel^{-1} \cdot h^{-1}$. The low molecular weight conjugates are disappearing at a rate of 135 $pmol \cdot kernel^{-1} \cdot h^{-1}$ and the IAA glucan is also lost at a rate of 135 $pmol \cdot kernel^{-1} \cdot h^{-1}$, and so this is some 2.7-fold faster than the measured rate of appearance of new IAA. Still this seems to be a reasonable agreement in view of the almost 10 year passage of time and the difference in seed batches used by Ueda[17] and by Epstein et al.[24] In fact, we have since found (Chisnell, personal communication) that there are large differences in the rates of endosperm softening depending upon the batch, variety and age of seeds.

The IAA being produced in the endosperm is just as quickly oxidized to oxindole-3-acetic acid (OxIAA) as recently shown by Reinecke.[25] Why should conjugates be hydrolyzed to form new IAA and the new IAA then converted to OxIAA at a rate such that the seed IAA pool remains steady state is totally unknown.

In any event, it is clear that there are two metabolic fates of the IAA conjugates in the *Zea* endosperm: 1) to be hydrolyzed and the IAA oxidized to OxIAA, and 2) for the conjugate to be transported from the endosperm into the shoot.

TRANSPORT OF IAA CONJUGATES FROM SEED TO SHOOT

Skoog first experimentally demonstrated the existence of a seed auxin precursor and, in fact, developed a quantitative assay for this substance.[26] His procedure was as follows: first an intact *Avena* seedling was decapitated and a block of agar placed on the cut stump. This block would presumably collect substances diffusing up from the seed into the shoot. Secondly, this block, following the collection period, was placed asymmetrically on a detipped and deseeded *Avena* plant in the conventional manner for the deseeded *Avena* test. Under these conditions, any free IAA in the block would produce curvature in the receptor plant within the customary 30 to 90 min. What was observed was that a curvature did not develop until several hours following block application. Skoog concluded that the block had collected the seed auxin precursor and that time was required for this precursor to be converted into free IAA and cause curvature in the receptor tissue. He found that tryptamine added to the agar block gave similar kinetics of curvature when placed on a receptor shoot. Guided then by that same strong interest in the conversion of tryptophan to IAA, he postulated that the seed auxin precursor was tryptamine or a related compound. Again, with the 20-20 vision of hindsight, it is easy to see why a known defined system was so much more attractive than an ill defined IAA complex.

To the best of our knowledge no one has subsequently studied the Skoog experiment. Using currently available methods, it should be possible to find out what went into the block and whether the time lag measured the length of time required to hydrolyze the seed auxin precursor to IAA.

The work of Nowacki and Bandurski[27] and, more recently, that of Chisnell (personal communication) demonstrates unequivocally that IAA-*myo*-inositol can function as a seed auxin precursor. Here the proof is as follows: 1) IAA-*myo*-inositol is a major seed auxin conjugate (cf. Ref. 1); 2) IAA-*myo*-inositol also occurs in the vegetative shoot;[28] 3) application of labeled IAA-*myo*-inositol to the endosperm results in IAA-*myo*-inositol and free IAA in the shoot;[27] and 4) there is an enzyme in the shoot which hydrolyzes IAA-*myo*-inositol to free IAA.[29]

There is ample documentation of the occurrence and quantitative estimation of IAA-*myo*-inositol in seed tissue (cf. Ref. 1) and so this will not be further considered here. Point 2 above, that is the occurrence of IAA-*myo*-inositol in shoot tissue, has only recently been documented and merits further discussion here. By using ^{3}H-labeled IAA-*myo*-inositol as a tracer and as an internal standard, Chisnell was able to demonstrate that IAA-*myo*-inositol constitutes at least 12% of the total IAA of vegetative tissue of *Zea*, a figure not too far from the 18% it comprises of the endosperm reserves. Further, owing to the difficulty of simultaneously isolating and purifying all four of the chemically separable isomeric IAA-*myo*-inositols, it is certain that this 12% is a minimum figure. In any event, it has been established that IAA-*myo*-inositol occurs in both seed and vegetative tissue of *Zea*.

Now the demonstration that IAA-*myo*-inositol applied to the endosperm appears in the shoot as free IAA, as IAA-*myo*-

inositol, and as other esters have been made by Nowacki and Bandurski and by Chisnell (personal communication). A quantitative summary of Nowacki's data is presented in Table 3. As can be seen some 7% of the transported radioactivity from IAA-*myo*-inositol applied to the endosperm appears as free IAA in the shoot. Almost 50% remained as IAA-*myo*-inositol and some 30% was present as esters other than IAA-*myo*-inositol. Chisnell obtained even more extensive hydrolysis of IAA-*myo*-inositol to free IAA and could, in fact, demonstrate that essentially *none* of the IAA of the coleoptile originated as free IAA in the endosperm.

The role of tryptophan as a seed auxin precursor has been discussed above and has recently been reviewed.[2-4] It should be said though that an exact quantitative assessment of the contribution of tryptophan in the endosperm to IAA in the shoot is extremely difficult. The concentration of free tryptophan in both endosperm and shoot is, at least, 10 times higher than that of IAA[24] and so even a few dpm of tryptophan converted to IAA would appear as a large conversion when corrected for pool size. This is further complicated by the facile non-enzymatic conversion of tryptophan to IAA.[24] We have studied this problem[27,30] with the conclusion that the conversion of tryptophan to IAA is small relative to the rate at which IAA-*myo*-inositol is hydrolyzed to IAA (Table 4).

Table 3. Distribution of radioactivity in shoot after application of [^{14}C]-IAA-inositol to the endosperm of a *Z. mays* seedling.[27]

Parameter	Amount pmol·shoot^{-1}·h^{-1}	Total %
Radioactivity in shoot	6.8	100
IAA plus ester IAA	6.3	93
IAA	0.5	7
Ester IAA	5.8	85
IAA-*myo*-inositol	3.8	56
Esters other than IAA-*myo*-inositol	2.0	29
Radioactivity not accounted for and presumably used in growth	0.5	8

Table 4. Transport of IAA, tryptophan-derived IAA, or IAA-*myo*-inositol from endosperm to shoot of *Z. mays*.[27]

Basis for calculation	$pmol \cdot shoot^{-1} \cdot h^{-1}$
[^{3}H]IAA applied to endosperm and appearing as [^{3}H]-IAA	0.015
[^{3}H]Tryptophan applied to endosperm and appear as [^{3}H]IAA in shoot.	0.15
[^{14}C]IAA-*myo*-inositol applied to endosperm and appearing as [^{14}C]IAA or [^{14}C]IAA esters in the shoot.	6.3
Amount of IAA needed by shoot.	5 to 9

The problem though should be restudied using lower specific activity tryptophan so as to avoid large multipliers.

Momonoki and Bandurski have recently initiated a series of studies of the effect of deseeding young *Zea* plants.[31] The experiments become complicated owing to the fact that the growth rate of the shoot changes subsequent to removal of the kernel, and thus, one does not know whether the observed reduction of hormone levels change owing to the reduced growth rate of the shoot or whether shoot growth rate changes owing to deprivation of the hormone or some other factor. Nonetheless, these data show that the shoot of an intact seedling is increasing in total IAA (ester plus free) at the rate of about 2 $pmol \cdot shoot^{-1} \cdot h^{-1}$. Deseeding a plant on the fourth day causes a small loss in total IAA and a relatively large free IAA loss. A portion of these results, adapted from this study, are summarized in Figure 1.

It is difficult to make conclusions. Data of Yomo and Varner[32] show that removal of the shoot stops the increase in protease activity of the cotyledons. Dr. Momonoki's data show that removal of the seed causes substantial reduction in shoot growth rate and in the free IAA content of the shoot. Somehow the shoot makes demands upon endosperm reserves and the endosperm, in turn, limits shoot growth. A parallel situation occurs in the case of fruit growth. If a fruit is present and growing, it makes demands upon the parent plant. If the fruit is removed the demand on the parent plant ceases, and, in some plants, photosynthetic rates are actually suppressed. We have only the dimmest notions of this two way system of supply and demand. One thing seems certain though, and that

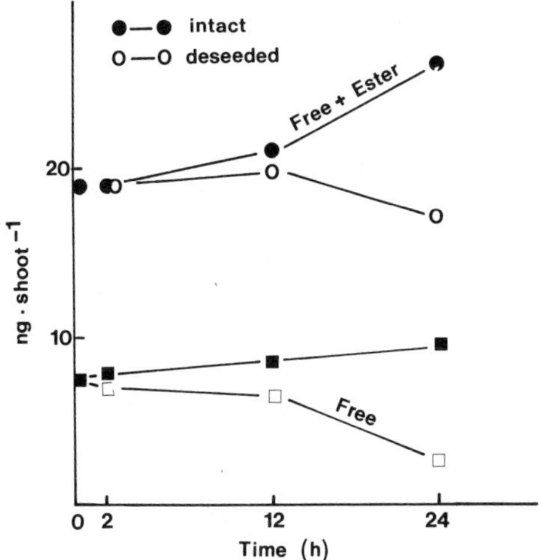

Fig. 1. Effect of removing the entire kernel, including scutellum, from 4 day old dark grown seedlings of Zea mays on IAA content of the shoot, as a function of time after deseeding. Data of Y. Momonaki. Open symbols are deseeded at zero h.

is that demand side economics are more effective than the supply side economics so recently studied in U.S. economics.

A further example of this two way communication lies in the case of the role of the tip of a young seedling shoot in controlling growth. If the tip of the plant is removed then growth ceases or, at least, is transferred to another portion of the plant. The tip must have been receiving a nutrient substance or hormone precursor from the seed and when the tip is removed the supply ceases or is transferred elsewhere.

We cannot answer these questions but wish only to indicate that the analytical capabilities that are becoming available should permit untangling the supply and demand situation posed by the shoot and the seed.

CONDUCTIVE TISSUES INVOLVED IN SEED AUXIN TRANSPORT

This discussion is moving from areas of relative certainty, including the chemistry of the conjugates and their rates of hydrolysis to areas of almost complete mystery including what determines the rates of hydrolysis and now, lastly, the transport tissues involved. In this case only two things are certain: first, that free IAA occurs mainly in the stele whereas ester IAA occurs mainly in the cortex.[33] This greater concentration of free IAA in the stele was first shown for roots by Greenwood et al[34] and by Bridges et al[35] and we have now extended it to shoots.[33] The ester IAA measurements had not previously been made and show that the pool of ester IAA is in the cortex.

A large body of knowledge indicates that free IAA from the apex moves downward through the phloem, cambium and derivative cells.[36] Our data demonstrate that IAA-*myo*-inositol is diffusing upward as a source of free IAA[27] and we had expected that the ester IAA would occur in the stele and this proved not to be the case. It will be necessary to determine in which tissue the ester is moving and where ester hydrolysis occurs. It is possible that the ester leaks out from the stele along its entire length and at each point an equilibrium between free and ester IAA is established or, alternatively, ester hydrolysis could occur only at the tip. Indeed Hall and Bandurski have found that the specific activity of the enzyme hydrolyzing IAA-*myo*-inositol to yield free IAA is higher in the coleoptile than in the mesocotyl.[29] Further, we need kinetic data for the shoot - IAA pools and knowledge of the rate of turnover of conjugated and free IAA in both cortical and stele tissue.

SUMMARY

Our summary is couched in terms of the working hypothesis illustrated in Figure 2. First, the hormone leaves the seed in the form of an IAA conjugate--in the case of *Zea* this is IAA-*myo*-inositol. Secondly, the conjugate is hydrolyzed in the stele. Thirdly, there is a controlled and variable rate of leakage of IAA from the stele into the cortex. Fourthly, in the cortex there is again established an equilibrium between free and ester IAA. The free IAA then reacts with a binding site and there commits the growth promoting act. Following growth promotion the IAA is oxidatively converted to OxIAA. Lastly, environmental stimuli control both the rate of leakage and the subsequent equilibrium between free and conjugated IAA. Thus, in the final analysis it is all of the sources of IAA including transport from the seed to the shoot, conjugate synthesis and hydrolysis, *de novo* synthesis,

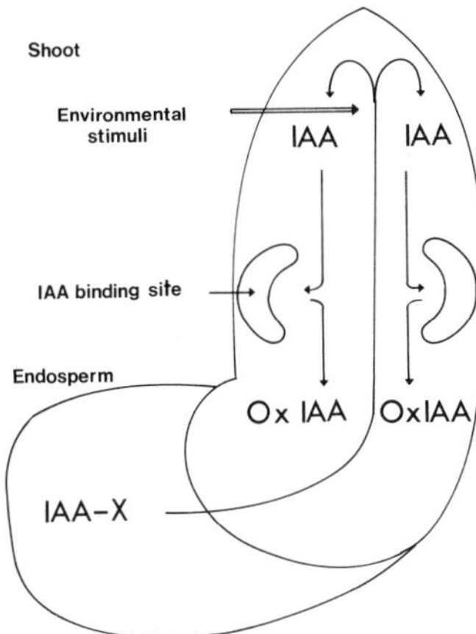

Fig. 2. A working hypothesis for the mobilization of seed auxin reserves during germination. IAA conjugates, designated IAA-X, leave the endosperm, permeate the scutellum, and enter the vascular tissue. Where hydrolysis of IAA-X to IAA occurs is unknown. Enzymes of the shoot, which are controlled by environmental stimuli, maintain a balance between IAA and IAA-X. Free IAA diffuses from the coleoptile tip and the mesocotyl node downward to the growing regions. Following growth promotion the IAA is "destroyed" by conversion to OxIAA.

and the rate of "use" of IAA which control early seedling growth.

ACKNOWLEDGMENTS

We acknowledge with pleasure the continued support of the Metabolic Biology Section of the National Science Foundation (PCM 79-04637) and the Life Science Section of the National Aeronautics and Space Administration (NAGW-97, ORD 25796). In addition, I am grateful for the use of, as yet, unpublished data of J. Chisnell, P. J. Hall, Y. Momonoki, D. Reinecke and A. Schulze and for assistance in manuscript preparation by M. D. La Haine. This is journal article No. 10525 of the Michigan Agricultural Experiment Station.

REFERENCES

1. COHEN JD, RS BANDURSKI 1982 Chemistry and physiology of the bound auxins. Annu Rev Plant Physiol 33: 403-430
2. SEMBDNER G, D GROSS, HW LIEBISCH, G SCHNEIDER 1980 Biosynthesis and metabolism of plant hormones. In J MacMillan, ed, Hormonal Regulation of Development I, Encyclopedia of Plant Physiology, New Series, Vol 9, Springer-Verlag, Berlin pp 281-444
3. SCHNEIDER EA, F WIGHTMAN 1974 Metabolism of auxin in higher plants. Annu Rev Plant Physiol 25: 487-513
4. SCHNEIDER EA, F WIGHTMAN 1978 Auxins. *In* DS Lethan, PB Goodwin, TJV Higgins, eds, Phytohormones and Related Compounds. A Comprehensive Treatise, Vol 1, Elsevier/North Holland, Amsterdam pp 29-105
5. CHOLODNY NG 1935 Über das keimungshormon von gramineen. Planta 23: 289-312
6. BERGER J, GS AVERY 1944 Chemical and physiological properties of maize auxin precursor. Am J Bot 31: 203-208
7. KÖGL F, AJ HAAGEN-SMIT, H ERXLEBEN 1934 Über ein neues auxin (heteroauxin) aus harn. Hoppe-Seyler's Z. Physiol Chem 228: 90-103
8. THIMANN KV 1935 On the plant growth hormone produced by *Rhizopus suinus*. J Biol Chem 109: 279-291
9. WILDMAN SG, MG FERRI, J BONNER 1947 An enzymatic conversion of tryptophan to auxin by spinach leaves. Arch Biochem 13: 131-146
10. SCHOCKEN V 1949 Genesis of auxin during decomposition of proteins. Arch Biochem 23: 198-204
11. LABARCA C, PB NICHOLLS, RS BANDURSKI 1965 A partial characterization of indoleacetylinositols from *Zea mays*. Biochem Biophys Res Comm 20: 641-646
12. LOEWUS FA, DB DICKINSON 1982 Cyclitols. *In* FA Loewus, W Tanner, eds, Plant Carbohydrates I, Encyclopedia of Plant Physiology, New Series, Vol 13A, Springer-Verlag, Berlin pp 193-216
13. AGRANOFF BW, JE BLEASDALE 1978 The acetylcholine phospholipid effect: What has it told us? What is it trying to tell us? *In* WW Wells and F Eisenberg Jr, eds, Cyclitols and Phosphoinositides, Academic, New York pp 105-120
14. HALL PJ 1980 Indole-3-acetyl-*myo*-inositol in kernels of *Oryza sativa*. Phytochemistry 19: 2121-2122

15. PISKORNIK Z, RS BANDURSKI 1972 Purification and partial characterization of a glucan containing indole-3-acetic acid. Plant Physiol 50: 176-182
16. PISKORNIK Z 1975 Distribution of bound auxins in kernels of sweet corn (Zea mays L.). Acta Biol Cracov 18: 1-12
17. UEDA M, RS BANDURSKI 1969 A quantitative estimation of alkali-labile indole-3-acetic acid compounds in dormant and germinating maize kernels. Plant Physiol 44: 1175-1181
18. PERCIVAL FW, RS BANDURSKI 1976 Esters of indole-3-acetic acid from Avena seeds. Plant Physiol 58: 60-67
19. COHEN JD 1982 Configuration of the aspartate moiety of indole-3-acetyl-aspartate isolated from soybean. Plant Physiol 69(Suppl): 12
20. BANDURSKI RS, A SCHULZE 1977 The concentration of indole-3-acetic acid and its derivatives in plants. Plant Physiol 60: 211-213
21. ZENK MH 1962 Aufnahme und stoffwechsel von α-naphthyl-essigsäure durch erbsenepicotyle. Planta 58: 75-94
22. HANGARTER RP, MD PETERSON, NE GOOD 1980 Biological activities of indoleacetylamino acids and their use as auxins in tissue culture. Plant Physiol 65: 761-767
23. HATCHER ESJ 1943 Auxin production during development of the grain in cereals. Nature 151: 278-279
24. EPSTEIN E, JD COHEN, RS BANDURSKI 1980 Concentration and metabolic turnover of indole in germinating kernel of Zea mays L. Plant Physiol 65: 415-421
25. REINECKE DM, RS BANDURSKI 1981 Metabolic conversion of ^{14}C-indole-3-acetic acid to ^{14}C-oxindole-3-acetic acid. Biochem Biophys Res Comm 103: 429-433
26. SKOOG F 1937 A deseeded Avena test method for small amounts of auxin and auxin precursors. J Gen Physiol 20: 311-334
27. NOWACKI J, RS BANDURSKI 1980 myo-Inositol esters of indole-3-acetic acid as seed auxin precursors of Zea mays L. Plant Physiol 65: 422-427
28. CHISNELL JR, RS BANDURSKI 1982 Isolation and characterization of indol-3-yl-acetyl-myo-inositol from vegetative tissue of Zea mays. Plant Physiol 69(Suppl): 55
29. HALL PJ, RS BANDURSKI 1981 Hydrolysis of ^{3}H-IAA-myo-inositol by extracts of Zea mays. Plant Physiol 67(Suppl): 2
30. HALL PL 1978 Movement of indole-3-acetic acid and tryptophan-derived indole-3-acetic acid from the

endosperm to the shoot of Zea mays L. Plant Physiol 61: 425-429

31. MOMONOKI Y, RS BANDURSKI 1982 Increase of indol-3-yl-acetic acid in germinating maize. Plant Physiol 69(Suppl): 12

32. YOMO H, JE VARNER 1972 Control of the formation of amylases and proteases in the cotyledon of germinating peas. Plant Physiol 51: 708-715

33. PENGELLY WL, PJ HALL, A SCHULZE, RS BANDURSKI 1982 Distribution of free and ester indole-3-acetic acid in the cortex and stele of the Zea mays mesocotyl. Plant Physiol 69: 1304-1307

34. GREENWOOD MS, JR HILLMAN, S SHAW, MB WILKINS 1973 Localization and identification of auxin in roots of Zea mays. Planta 109: 369-374

35. BRIDGES IG, JR HILLMAN, MB WILKINS 1973 Identification and localization of auxin in primary roots of Zea mays by mass spectrometry. Planta 115: 189-192

36. GOLDSMITH MHM 1977 The polar transport of auxin. Annu Rev Plant Physiol 28: 439-478

Chapter Twelve

EFFECT OF GERMINATION ON CEREAL AND LEGUME NUTRIENT CHANGES AND FOOD OR FEED VALUE: A COMPREHENSIVE REVIEW

P. L. Finney

Western Wheat Quality Laboratory
Agricultural Research Service
United States Department of Agriculture
and
Department of Food Science
Washington State University
Pullman, Washington 99164

INTRODUCTION

Cereals and legumes are the foodstuffs for most humans and animals and have been throughout recorded history. To extract "maximum nutrients for minimum costs," the seeds of those plants have usually been treated by germinating, fermenting or selectively heat treating to increase the amount or availability of nutrients.

In the so-called Western and highly industrial world the practice of sprouting is largely limited to malting cereals for the brewing industry; but for hundreds of millions of other world citizens, sprouting of legumes and some cereals is routine in converting feed grains into human foods. However, for the past century researchers in both the Western- and Eastern-world have studied the effects of germination on the physical, physiological, biochemical, nutritional, and food functional properties of cereals and legumes.

This chapter brings together data from those studies and evaluates the ancient practice of sprouting seeds for food and feed uses from a contemporary viewpoint. Consolidation of this information about seed sprouting from numerous disciplines provides a view of the role that germination studies have played in the evolution of biological chemistry and how intimately interrelated it is with food science and nutrition.

VITAMIN C

Human and Animal Feeding Studies

It was during the first decade of this century that Fürst[1] established that both barley and lentils when germinated produced substantial quantities of antiscorbutic materials. Predating Fürst's work by as much as 400 years are numerous naval records which document the superiority of fresh versus dehydrated, germinated barley- and wheat-wort beers in preventing and curing scurvy among sailors.[2] Even Captain Cook in his famous voyages of discovery, after being at sea from November 22, 1772 to March 26, 1773 attributed the lack of scurvy aboard ship to use of fresh wort beer and sweet wort as a prophylaxis and curative agent against scurvy.[2] Chick and Hume[3] confirmed Fürst's work and concluded that germinated legumes and cereals "occupy a special position among foodstuffs in being richly endowed with both the anti-scurvy and anti-beri-beri vitamin."

Deaths due to disease and malnutrition during World War I prompted studies to establish which foods could serve troops in combat to eliminate scurvy. Grieg[4] in India confirmed the antiscorbutic value of sprouted peas, lentils, chickpeas, blackeye peas and black gram. Chick et al[5] and Chick and Delf[6] established that 5 g germinated peas or lentils possessed the antiscorbutic potency of 5 g of fresh beans and approximately equaled 1.5 ml of either lemon or orange juice and was more potent than 20 g carrot in guinea pig feeding trials. In human feeding experiments conducted in a military infirmary, Wiltshire[7] fed 27 severe cases of scurvy 4 oz per day of 3 day germinated haricot beans and fed 30 other patients 4 oz fresh lemon juice. Beans outperformed the juice with the result that 70% versus 53% of the cases were cured within 4 weeks.

Dyke[8] discovered that South African natives developed scurvy when their native beer, which had previously been produced from germinated maize or millet, was replaced by beer produced from meal of ungerminated maize or millet. Immediately thereafter Harden and Zilva[283] showed using guinea pigs that many common beers and ales lacked vitamin C activity since they too were being produced from ungerminated cereal meals and relatively small amounts of high-amylase malts. Harden and Zilva[9] confirmed those results using monkeys which developed scurvy when fed beers but grew well on fresh or dehydrated barley malts.

Those studies prompted work on the evaluation of local food grains for vitamin C throughout the world for the next 3 to 4 decades. Delf[10] found 3 day germinated cowpea protected guinea pigs from scurvy for 90 days. Santos[11] fed 5 g of mung bean sprouts each day and kept guinea pigs alive without signs of scurvy for 90 days. Delf[10] soaked and germinated cowpeas, soybeans and sorghum at 25° to 30°C for 3 to 5 days. As with previous experiments with clipper peas in England, he found a ration of 2.5 g raw, germinated cowpeas protected guinea pigs from scurvy for nearly 90 days. Sorghum had only slight antiscorbutic effects while the soybeans had no detectable vitamin C when germinated in a similar manner. Honeywell and Steenbock[12] reported the concentration of vitamin C increased substantially in germinating barley, which happened even in the dark. Vitamin C production did not occur during the first 24 h soaking nor was it produced in the absence of oxygen even if soaked 96 h. Kucera[284] reported that germinating rye for only 24 h produced enough vitamin C to prevent scurvy when fed (presumably) ad libitum while barley and wheat required about 4 days of germination to have an antiscorbutic effect when fed to guinea pigs. Oats required somewhat more germination time to reach a comparable vitamin C level. Matsuoka[310] reported 2.5 g of rice germinated 3 days was sufficient to prevent scurvy in guinea pigs. Simonik[311] found measurable amounts of vitamin C after 6 h germination of vetch, broadbeans, and peas. After 4 days of germination, only 2 g/day maintained guinea pigs with no signs of scurvy. A quarter century later French et al[13] soaked peas, cowpeas, soybeans, and navy-, baby lima-, red kidney- and pink-beans and emphasized prolonged soaking (up to 24 h) only accelerated appearance of rot. Wu[14] indicated germinated soybeans were an excellent source of vitamin C.

Miller and Hair[15] found that mung bean sprouts compared favorably in vitamin C activity to lemon-, orange-, and tomato-juice in guinea pig studies. Wats and Eyles[16] reported that mung beans were better producers of vitamin C activity and easier to germinate than chickpeas, peas, black gram, lobia, and soybeans. Five g of mung bean sprouts fed daily prevented scurvy in guinea pigs for 80 days. Bogart and Hughes[17] found the vitamin C content in germinating oats increased up to the tenth day and reported 40 g (wet wt) of the 7 day sprouts fed daily to guinea pigs was adequate as the sole source of vitamin C. Ghosh and Guha[18] demonstrated that germination of black gram produced sufficient vitamin C to avoid scurvy in guinea pigs with 4 g/day (dry wt) for 68 days.

Kahn[19] in describing the Indian famine of 1938-1941, stated that "in December 1939 a rather sudden deterioration in health conditions occurred in the district, accompanied by a material increase in death rates. This led to careful investigations in a number of selected villages which resulted in the discovery of cases of scurvy. From December 1939 on to the end of March 1940, cases continued to occur, but with the application of appropriate preventive measures the disease disappeared during the months of April, May and June, 1940, only to reappear when those measures were in the first instance curtailed and later discontinued altogether." Scurvy and malnutrition were essentially eliminated by the issue of germinated grain. In fact, when the germinated grain was discontinued, the disease reappeared, and thereafter disappeared when the germinated grain was reintroduced as a prophlyactic. As a curative and preventative measure over 200,000 people received 1 oz germinated grain biweekly.

Chemical Analyses for Vitamin C in Sprouts

By the early 1930's chemical approximation of vitamin C content in food grains became more popular and more reliable. One key to that improvement, discovered by Johnson,[20] was that indophenol-reducing capacity could not be equated with vitamin C content or with antiscorbutic activity. Ray[21] analyzed only the sprouted portion of peas which increased from nil to about 50 mg/100 g (wet wt) after 90 h germination.

By 1935 Ahmed[22,23] reevaluated previous data and estimated chemical quantities of vitamin C from the reported biological values. He noted excellent agreement between chemical and biological tests for vitamin C from his and earlier studies. Lee[24] found peas sprouted in the dark contained about 40 mg/100 g on the eighth day. Subsequently, most common, edible cereals and legumes were analyzed for vitamin C increases due to germination (Table 1).[25-37] The vitamin C content of dry ungerminated cereals and legumes is generally close to zero, although some legumes may contain a few milligrams per 100 g (dry basis). Two notable exceptions have been documented. Bhagvat and Rao[38] reported dry, unsprouted bengal gram contained sufficient vitamin C to promote good growth in guinea pigs when included in the diet at a 32% level. Also Hamilton and Vanderstoep[39] found 48 mg ascorbic acid/100 g, (dry basis) of unsprouted alfalfa.

Probably varietal and environmental variations in all seed species affect vitamin C content and other seed constituents. For example, Fordham et al[34] showed different sprouted pea varieties ranged in vitamin C content from 18.8 to 50 mg/100 g (dry basis) while various bean varieties ranged from 12.6 to 42.2 mg/100 g (dry basis). Many varietal and environmental differences in seed nutrients are not evident in this review since the tables report mean values when more than one variety or seed lot is reported by a given source.

Effect of Light on Vitamin C Synthesis

There are numerous reports in which the effect on vitamin C production by various degrees of sunlight were evaluated during early seed germination. As early as 1927 Heller[40] fed guinea pigs yellow milo maize or wheat germinated in the dark or in sunlight and noted the animals gained almost twice as much weight on sprouts grown in sunlight. Death or scurvy resulted from feeding sprouts grown in the dark to other animals. Matsuoka[285] reported that to avoid scurvy guinea pigs required about 1/4 as much extract from germinated rye or rice if the seeds were germinated in the light. Others have corroborated Heller's biological studies with analytical studies[39,41-45] (Table 2).

Kucera[286] reported 6 to 12 day germinated rye seed, root or stem all contained nearly equal vitamin C activity as evaluated by guinea pig feeding. Sunlight had no effect on

Table 1. Vitamin C content of some edible seeds and sprouts.

Common Name[a]	Scientific Name	Germ. Time (Days)	Vitamin C (mg/100 g)[b]	Reference Number
Pea	*Pisum sativum*	4	86	25
Oat	*Avena byzantina*	5	42	
Barley	*Hordeum vulgare*	2	23	26
Mustard	*Brassica nigra*	4	65	27
Sesame	*Sesamum indicum*	4	30	
Rice	*Oryza sative*	4	6	
Barley	*Hordeum vulgare*	4	10	
Wheat	*Triticum vulgare*	4	13	
Sorghum	*Sorghum vulgare*	4	15	
Maize	*Zea mays*	4	15	
Millet	*Pennisetum glaucum*	?	?,14	28
Maize	*Zea mays*	?	16,14	
Wheat	*Triticum vulgare*	?	18,30	
Sorghum	*Sorghum vulgare*	?	42,29	
Cowpea	*Vigna cylindrica*	?	37,25	
Black soybean	*Glycine max*	?	64,47	
Green soybean	*Glycine max*	?	27,18	
Barley	*Hordeum vulgare*	3	20	29
Wheat	*Triticum vulgare*	3	17	
Chickpea	*Cicer arietinum*	3	45	
Mung bean	*Phaseolus aureus*	3	124	
Wheat	*Triticum vulgare*	4	15	30
Maize	*Zea mays*	4	11	
Rice	*Oryza sativa*	4	5	
Millet	*Pennisetum glaucum*	4	24	
Sorghum	*Sorghum vulgare*	4	16	
Khesari dahl	*Lathyrus sativus*	2	27	31
Bengal gram	*Cicer arietinum*	2	33	
Green gram	*Phaseolus radiatus*	2	53	
Cowpea	*Vigna cylindrica*	2	47	
Soya bean	*Glycine hispida*	2	11	31
Pea	*Pisum sativum*	2	48	
Wheat	*Triticum vulgare*	2	9	
Jowar	*Sorghum vulgare*	2	5	

Table 1 continued.

Common Name	Scientific Name	Germ. Time (Days)	Vitamin C (mg/100 g)	Reference Number
Sona mung(3)	*Phaseolus radiatus*	3	96	32
Kash kalai	*Phaseolus mungo*	4	81	
Kurathi kalai	*Dolichos uniflorius*	4	79	
Motor(3)	*Pisum sativum*	5	63	
Chola(3)	*Cicer arietinum*	4	41	
Khesarii	*Lathyrus sativus*	4	25	
Mashuri	*Lens esculenta*	5	87	
Barbati	*Vigna sinensis*	5	55	
Soyabean	*Glycine max*	5	61	
French bean	*Dolichos sp-red*	5	24	
French bean	*Dolichos sp-white*	5	46	
Chim	*Dolichos lablab*	5	35	
Arahar	*Cajanus indicus*	5	29	
Joar	*Sorghum vulgare*	5	26	
Soybean	*Glycine max*	6	104	33
Pea(6)	*Pisum sativum*	5	38	34
Bean(6)	*Phaseolus vulgarus*	5	28	
Pinto	*Phaseolus vulgarus*	5	39	
Red Kidney	*Phaseolus vulgarus*	5	13	
Red valentine	*Phaseolus vulgarus*	5	40	
Navy	*Phaseolus vulgarus*	5	22	
Soybean	*Glycine max*	5	21	
Mung	*Phaseolus aureus*	5	38	
Black gram	*Phaseolus mungo*	3+	18	35
Soybean	*Glycine max*	3	25	36
Yellow pea	*Pisum sativum*	4	64	37
Lentil	*Lens esculenta*	4	78	
Faba bean	*Vicia faba*	4	76	

[a] Number of varieties tested in parantheses for all tables.

[b] Dry basis reported in all tables unless otherwise stated.

Table 2. Effect of sunlight on vitamin C production in some cereal and legume sprouts.

Common Name	Scientific Name	Germ. Time (Days)	Vitamin C Dark (mg/100 g)	Vitamin C Light (mg/100 g)	Gain From Light %	Reference Number
Oat	*Avena byzantina*	10	11[a]	11[a]	-	17
Soybean	*Glycine max*	9,14	2.3[b]	3.3[b]	+43	41
Oat	*Avena byzantina*	3	46	75	+63	42
Cowpea	*Vigna cylindrica*	4	16	30	+88	43
Soyabean	*Glycine max*	5	17	35	+106	44
Chickpea	*Cicer arietinum*	6	110	186	+69	45
Mung bean	*Phaseolus aureus*	5	190	326	+72	
Soybean	*Glycine max*	5	72	91	+26	
Pea	*Pisum sativum*	3	25	26	+2	47
Mung bean	*Phaseolus aureus*	5	62	63	+1	
Cowpea	*Vigna cylindrica*	6	54	53	-2	
Black gram	*Phaseolus mungo*	3	40	42	+5	
Alfalfa	*Medicago sativa*	3	73	126	+73	39
Alfalfa	*Medicago sativa*	5	106	178	+70	
Mung bean	*Phaseolus aureus*	2.5	132	145	+10	48

[a] mg/500 seedlings. [b] Coleoptile only.

vitamin C production. Roy et al[46] indicated that light may not promote vitamin C synthesis. Although indophenol reducing substances in soybeans and mung beans increased more in the presence of light than in the dark the reducing substances were not identical with ascorbic acid as evidenced by preliminary bio-assays with guinea pigs. Others,[17,47,48] also reported sunlight had no effect on vitamin C production during germination (Table 2).

What are the reasons for differences in data on the effect of sunlight on vitamin C production in sprouting edible seeds? Certainly different seed species, and different seed lots depending on their growing conditions, respond as contrarily as the data in Table 2 indicate. Some differences are attributable to differences in seed species, growing conditions, germination procedures, and degree of hydration prior to prolonged germination. In addition, sunlight may promote vitamin C production when mineral salts are absent from the soaking and germination medium. Without a sufficient mineral balance in the steep water, relatively large amounts of mineral salts are leached from the seeds, creating suboptimal conditions for vitamin C synthesis. Among those who found vitamin C production was enhanced by sunlight (Table 2), only Reid[43] seems to have germinated seeds in tap water. Among those studies in which sunlight reportedly had no bearing on vitamin C production, two involved the use of tap water. Also, De and Barai[47] soaked seeds in Knopps solution (0.8 g $Ca(NO_3)_2$, 0.2 g KNO_3, 0.2 g KH_2PO_4 , 0.2 g $MgSO_4$, and a trace amount of $FePO_4$/10 L).

Effect of Manganese on Vitamin C Synthesis

Rudra[49] soaked bengal gram (chickpea) in 0, 1, 10, or 100 ppm $MnSO_4$ solution, and reported vitamin C contents of 64, 76, 87 and 85 mg/100 g (dry basis), respectively, after 6 days germination. Growth was enhanced in 1 and 10 ppm $MnSO_4$ solution and repressed in 100 ppm when compared to the distilled water control. Rudra[50] confirmed that Mn is a determining factor in *in vitro* ascorbic acid synthesis in rat tissue. $MnCl_2$ enhanced vitamin C production from galactose, like mannose and glucose, but to a less extent in rat liver than in guinea pig liver. Rudra[51] also reported guinea pig liver synthesized ascorbic acid precursors both *in vitro* and *in vivo*. The concentration of Mn necessary for synthesis of guinea pig ascorbic acid was about 10 to 100 times greater than in the case of the rat liver. Rudra[51] hypothesized that the inability of the guinea pig and man to synthesize ascorbic acid is due to inadequate Mn in their tissues.

In agronomic studies Hester[52] found additions of $MnSO_4$ to soil materially increased vitamin C production in tomatoes. Ahmed et al[45] evaluated the effects of adding from 0.005 to 0.01% of $CoCl_2$, H_3BO_3, $MnCl_2$ or CaF_2 on vitamin C synthesis in green gram embryo cultures. Concentrations of 0.01% and 0.02% $MnCl_2$ and 0.01% H_3BO_3 stimulated ascorbic acid production, however higher concentrations of those mineral ions reduced ascorbic acid synthesis, as did CaF_2 at lower concentrations. De and Barai[47] studied the effects of adding 0, 10, 100, or 1000 ppm $Fe(NO_3)_3$, $Cu(NO_3)_2$, or $MnCl_2$ on the biosynthesis of ascorbic acid during 48 h germination of mung beans in light. They found 10 ppm was optimum and increased vitamin C content from about 43 to 59, 65 and 79 mg/100 g (dry basis) respectively for the three salts. A second study by the same authors confirmed those results and 15 ppm $MnCl_2$ further increased vitamin C to 82 mg/100 g (dry basis).

Effect of Sugars on Vitamin C Synthesis

Ray[21] reported the ascorbic acid producing potential of most hexoses varied between 19 and 32 mg/100 g. Mannose was most effective in Ray's excised pea germ studies, producing 62 mg of vitamin C/100 g. Among the sugars tested, xylose produced the most growth. Guha and Ghosh[53] found that minced rat kidney and spleen produced far more vitamin C from mannose than from any of the other 5 hexoses added to a tissue culture. Minced rat spleen produced as much ascorbic acid as did minced rat liver and kidney, but was not able to convert any of the other hexoses tested to ascorbic acid. Guha and Ghosh[54] also isolated an enzyme system from germinating mung beans which converted mannose into ascorbic acid at pH 5.8 but not 7.4. A wide range of activity was noted among minced liver from rat, rabbit, pigeon, guinea pig (normal), guinea pig (scorbutic), and monkey to produce vitamin C from mannose, with relative values of +0.30, +0.04, +0.05, -0.03, -0.02, and -0.01 mg/g, respectively. Guha and Ghosh[55] confirmed _in vitro_ studies with _in vivo_ studies of conversion of mannose to vitamin C in which rats were sacrificed after sugar injections into organs. Glucose did not stimulate ascorbic acid synthesis as did mannose. Guha and Ghosh[56] reported aqueous extracts from mung beans converted mannose to vitamin C at pH 5.8 in air but not in N_2 atmosphere.

Virtanen and Eerola[57] found the addition of glucose, fructose, mannose, or sucrose to steeping medium had no effect on vitamin C production in germinating peas. Reid[43] indicated that cowpeas germinated in the dark in 0.5% glucose approximately doubled vitamin C synthesis, compared to the control with no glucose. Rudra[50] confirmed that galactose may also act as precursor to ascorbic acid. In a continuation of earlier work, Rudra[51] reported that guinea pig liver also can synthesize vitamin C from mannose and galactose both _in vitro_ and _in vivo_. Ahmed et al[45] found a poor correlation between excised green gram embryo culture growth and the amount of vitamin C synthesized. Five percent glucose promoted the greatest vitamin C increase (392%), whereas xylose promoted both the least growth and the least vitamin C (15%) and maltose, mannose and lactose produced 104, 97 and 73% increases in vitamin C, respectively.

De and Barai[47] cultured cowpea embryo seedlings in Knopp's mineral salt solutions with and without 5% sucrose, glucose, fructose, galactose, or mannose. Concentrations as low as 0.05% mannose resulted in production of as much vitamin C as was produced by other sugars when tested at concentrations of 5%. Mannose concentrations of 0.05, 2.5, and 5% resulted in 43, 63, and 68 mg vitamin C per 100 g (dry basis), respectively.

Effect of Vitamin C as a Growth Stimulator

One hundred to 500 ppm ascorbic acid added to sprouting wheat accelerated shoot length growth 25 to 30%, increased shoot weight 25 to 30% and increased root weight 50%.[58] At 2500 ppm, ascorbic acid caused a slight inhibitory effect on germination rate and a 24 to 45% reduction in growth and weight of seedlings. Oats were much less sensitive to addition of vitamin C, being neither greatly stimulated nor inhibited. Addition of ascorbic acid to peas grown in sterile agar and Hiltner's solution with $Ca(NO_3)_2$ (pH 5.5) increased dry weight of plants at all stages of growth up to "full bloom with small pods" and increased total vitamin C content in sprouts and plants by 25 to 78%.[59,60] The greatest difference in vitamin C content occurred during sprouting. Growth of oat, garden cress (*Lepidium sativum*) and mustard (*Brassica alba*) seedlings was stimulated when 10 to 100 ppm ascorbic acid was added at the beginning of steep or after one or two days germination.[61]

WATER-SOLUBLE B-COMPLEX VITAMINS

Thiamin Animal Feeding Studies

Embrey[62] concluded from feeding studies with mice that sprouted mung beans were approximately twice as good a source of "water-soluble B vitamin" as the orginal beans. Santos[11] studied *Phaseolus mungo* and "Togi," a Phillipino term for germinated bean which was prepared by soaking the beans 24 h and germinating them 2 days. Beans or sprouts were fed to rats deficient in vitamin B_1. Whereas 1 g of beans per day was required to restore growth in rats only 1/2 g of sprouts per day was sufficient, indicating that thiamin approximately doubled during the 3 days of germination. Bowman and Yee[63] injected 3 mg anti-beri-beri vitamin from mung beans redissolved in water and alcohol into polyneuritis-induced pigeons previously fed polished rice and water for two weeks. "Within an hour these pigeons were able to stand erect and showed normal movement of head." In earlier studies, Harden and Zilva[9] found that "when malt was given pigeons as the source of the anti-neuritic factor the birds grew but eventually developed polyneuritis." Other pigeons fed one of 5 other commercial malts which had been dried from 195° to 220°F showed no symptoms of polyneuritis after two months but malts were no more effective than were ungerminated barleys. Rat feedings also showed that the anti-neuritic properties of those five malts and the original barley were approximately equal, indicating germination had not changed the vitamin B_1 content. Rose and Phipard[64] found rats fed 14 day sprouted pea roots and shoots gained 5.7 versus 4.2 g per week for thiamin deficient rats fed ungerminated peas as controls.

During the past 4 to 5 decades numerous edible seeds and sprouts were analyzed for thiamin content (Table 3). Generally, thiamin content remains unchanged after a few days germination. However, Burkholder and McVeigh[69] noted increases when seeds were cultured in sand media, and Chattopadhyay et al[74] reported 50 to 100% increases in thiamin content after 3 to 4 days germination for all seeds tested (Table 3).

The thiamin content of edible seeds ranged from 1 to 13 μg/g, with oats, soybeans, and alfalfa containing the most. Except for the data of Chattopadhyay et al,[74] only a few seed species gained in thiamin due to germination while most remained unchanged or lost 10 to 20% thiamin during a few days germination. Early feeding studies with mung bean sprouts and bean seeds showed apparent doubling of thiamin content after germination for a few days while similar studies using barley showed no change in thiamin content due to sprouting.

Riboflavin

Rose and Phipard[64] found 14 day sprouted peas supported the same growth of riboflavin-depleted rats as twice the amount of the unsprouted peas, indicating that germination approximately doubled that vitamin. Lima beans contained about 50% as much riboflavin as the peas in that study.

The riboflavin content of numerous edible seeds and sprouts as determined during the past 50 years is listed in Table 4. Invariably sprouting increased riboflavin content, generally by a few hundred percent. Lee and Whitcomb[79] reported soybeans germinated for 5 days contained about 0.90 μg/g (wet wt).

Davis et al[76] noted that American malts contained 300 to 600% more riboflavin than ungerminated barley. Klatzkin et al[80] found riboflavin levels of commerical English malt varied widely in vitamin B_2 content. Those authors generally found 100% increases due to germination, but conceded that the barley normally used in England "contains a good deal of the precursor of riboflavin, which can be converted into the vitamin during the brewing of malt extract." In fact, other workers found substantial increases in riboflavin in beer resulting from the brewing process.[81-84]

Wai et al[73] found 54 h germination optimized appearance and edibility of soybeans. By then the riboflavin content had doubled, and since 2/3 of the vitamin still remained in cotyledons, the report recommended that "the entire bean should be eaten." Fordham et al[34] found that pea and bean sprouts, including mung, soy, white navy and others, contained enough riboflavin to supply 1/3 recommended daily allowance per serving, or about 1.2 mg/day. Kylen and McCready[75] assayed riboflavin in seeds and 3-1/2 day sprouts of alfalfa, lentils, mung beans and soybeans. They found substantial increases in alfalfa (6.2-17.9 μg/g), mung beans (5.2-12.8 μg/g), and soybeans (2.5-6.0 μg/g), but not in lentils.

Table 4 summarizes the riboflavin content of over 150 seed lots, including over 90 sprouted samples. Invariably, vitamin B_2 content increased during sprouting, with the average value increasing from 1.54 to 5.44 μg/g (dry basis). Samples of alfalfa, mung beans, and soybeans contained the most vitamin B_2, while alfalfa, mung bean, oat and wheat sprouts had higher than average riboflavin values. The average vitamin B_2 concentration of three oat seed lots was 0.8 μg/g before germination. After germination their average was 11.7 μg/g, an increase of 1460%. Numerous edible cereals and legumes contained from 6 to 10 μg/g of riboflavin after 4 days germination (Table 4).

Table 3. Thiamin content of some edible seeds and sprouts.

Common Name	Scientific Name	Germ. Time (Days)	Thiamin Content Seed (µg/g)	Thiamin Content Sprout (µg/g)	Reference Number
Rice	Oryza sative	0	4.1	-	65
Wheat	Triticum vulgare	0	4.3	-	
Bengal gram	Cicer arietinum	0	4.5	-	
Black gram	Phaselous mungo	0	4.1	-	
Green gram	Phaseolus aureus	0	4.5	-	
Lentil	Lens esculenta	0	4.1	-	
Red gram(3)	Cajanus indicus	0	2.6	-	
Brewers yeast(2)	Saccharomyces	-	29.5	-	
Soybean(6)	Glycine max	0	6.4[a]/9.0[b]	-	66
Lima bean(3)	Phaseolus lunalatus	0	4.3	-	67
Blackeye pea	Vigna cylindrica	0	5.9	-	
Soybean(5)	Glycine max	0	9.2	-	
Pea(2)	Pisum sativum	0	7.2	-	
Peanut (2)	Arachis hypogea	0	8.5	-	
Lentil	Lens esculenta	0	6.9	-	
Pea	Pisum sativum	0	7.6	-	68
Red bean	Phaseolus vulgarus	0	6.8	-	
Pink bean	Phaseolus vulgarus	0	5.3	-	
Lentil	Lens esculenta	0	2.8	-	
Barley	Hordeum vulgare	5	-	7.9	69
Corn	Zea mays	5	6.2	5.5	
Oat	Avena byzantina	5	10.0	11.5	
Soybean	Glycine max	5	10.7	9.6	
Large lima	Phaseolus lunalatus	5	6.7	5.0	
Small lima	Phaseolus lunalatus	5	4.5	6.2	
Greeneye pea	Vigna sinensis	5	11.0	12.0	
Mung bean	Phaseolus aureus	5	8.8	10.3	
Pea	Pisum sativum	5	7.2	9.2	
Corn	Zea mays	5	5.1	5.2	
Oat	Avena byzantina	5	11.4	11.5	
Wheat	Triticum vulgare	5	5.3	9.8	
Barley	Hordeum vulgare	5	4.8	9.0	
Blackeye pea	Vigna cylindrica	1.5	8.5	12.0	70
Lima bean	Phaseolus lunalatus	2	5.8	4.8	
Cottonseed	Gossypium ?	1.5	3.2	4.9	
Oat	Avena byzantina	5-6	11.3	12.2	71
Wheat	Triticum vulgare	5-6	7.0	9.0	
Barley	Hordeum vulgare	5-6	6.8	9.0	
Corn	Zea mays	5-6	5.5	5.1	

Table 3 continued.

Common Name	Scientific Name	Germ. Time (Days)	Thiamin Content Seed (μg/g)	Thiamin Content Sprout (μg/g)	Reference Number
Pea(2)	Pisum sativum	3	7.8	6.9	13
Blackeye pea	Vigna cylindrica	3	8.4	-	
Soybean(6)	Glycine max	3	10.0	7.9	
Soybean(7)	Glycine max	4	16.2	16.2	72
Mung bean	Phaseolus aureus	4	8.6	8.6	
Soybean	Glycine max	3	12.8	11.0	73
Soybean	Glycine max	3	14.0	13.4	
Ghasi mung	Phaseolus mungo	3	2.8	4.2	74
Sona mung	Phaseolus mungo	3	1.8	3.8	
Kalai	Phaseolus radiatus	4	2.6	4.8	
Mash kalai	Phaseolus roxburgi	4	2.5	4.7	
Kurathi kalai	Dolichos uniflorous	4	0.8	2.6	
Motor(3)	Pisum sativum	4	2.6	4.4	
Chola(3)	Cicer arietinum	3	4.8	6.0	
Khesarii	Lathyrus sativus	4	2.1	3.2	
Mushuri	Lens esculenta	4	2.6	3.5	
Barbati	Vigna sinensis	3	0.8	1.6	
Soyabean	Glycine soya	4	2.6	4.9	
French bean	Dolicus sp-white	3	1.6	2.6	
Chim	Dolicus lablab	3	1.2	1.8	
Arahar	Cajanus indicus	4	2.5	3.0	
Joar	Sorghum vulgare	3	1.3	2.6	
Alfalfa	Medicago sativa	3.5	11.6	12.0	75
Lentil	Lens esculenta	3.5	8.0	7.7	
Mung bean	Phaseolus aureus	3.5	7.8	9.9	
Soybean	Glycine max	3.5	13.0	11.9	

[a] Green. [b] Mature.

Niacin

Nicotinic acid was identified as the pellagra preventive vitamin by Elvehjem and co-workers.[85] In man pellagra is associated with nicotinic acid deficiency and is related to the fact that the vitamin occurs in cereals in a bound and nutritionally unavailable form.[86] The nature of bound niacin in cereals has been well studied since Kodicek[87,88] discovered

Table 4. Riboflavin content of some edible seeds and sprouts.

Common Name	Scientific Name	Germ. Time (Days)	Riboflavin Seed (μg/g)	Riboflavin Sprout (μg/g)	Reference Number
Soybean(immature)(6)	Glycine max	0	3.6	-	66
Soybean(mature)(6)	Glycine max	0	2.3	-	
Lima bean(3)	Phaseolus lunalatus	0	1.7	-	67
Blackeye pea	Vigna cylindrica	0	2.3	-	
Soybean(5)	Glycine max	0	3.3	-	
Pea(2)	Pisum sativum	0	1.7	-	
Peanut(2)	Arachis hypogea	0	1.6	-	
Lentil	Lens esculenta	0	2.6	-	
Pea(8)	Pisum sativum	0	3.2	-	68
Red bean(9)	Phaseolus vulgarus	0	2.6	-	
Pinto bean(5)	Phaseolus vulgarus	0	2.7	-	
Lentil(3)	Lens esculenta	0	3.0	-	
Wheat	Triticum vulgare	5	1.0	2.4	69
Corn	Zea mays	5	1.5	2.0	
Soybean	Glycine max	5	1.8	3.6	
Mung bean	Phaseolus aureus	5	1.2	2.6	
Barley	Hordeum vulgare	5	1.3	8.3	
Corn	Zea mays	5	1.2	3.0	
Oat	Avena byzantina	5	0.6	12.4	
Soybean	Glycine max	5	2.0	9.1	
Lima bean(2)	Phaseolus lunalatus	5	1.0	3.0	
Greeneye pea	Vigna cylindrica	5	1.8	9.7	
Mung bean	Phaseolus aureus	5	1.2	10.0	
Pea	Pisum sativum	5	0.7	7.3	
Wheat[a]	Triticum vulgare	5	1.0	13.0	
Corn[a]	Zea mays	5	2.0	8.7	
Wheat[b]	Triticum vulgare	5	1.5	5.2	
Corn[b]	Zea mays	5	1.2	4.0	
Oat[b]	Avena byzantina	5	1.0	11.0	
Barley	Hordeum vulgare	5	0.5	6.9	
Blackeye pea	Vigna cylindrica	2	1.5	3.9	70
Lima bean	Phaseolus lunalatus	2	1.4	1.4	
Cottonseed	Gossypium ?	2	2.3	2.8	
Oat	Avena byzantina	5.5	0.8	11.6	71
Wheat	Triticum vulgare	5.5	1.3	5.4	
Barley	Hordeum vulgare	5.5	0.9	7.2	
Corn	Zea mays	5.5	1.1	4.3	
Barley(7)	Hordeum vulgare	6	1.0	3.1	76
Wheat(5)	Triticum vulgare	6	0.9	2.9	

Table 4 continued.

Common Name	Scientific Name	Germ. Time (Days)	Riboflavin Seed (μg/g)	Riboflavin Sprout (μg/g)	Reference Number
Pea(3)	*Pisum sativum*	3	0.3	1.7	13
Soybean(6)	*Glycine max*	3	0.6	1.3	
Oat	*Avena byzantina*	5	0.03[c]	0.29[c]	77
Soybean(7)	*Glycine max*	4	3.6	5.6	72
Mung bean	*Phaseolus aureus*	4	3.0	14.4	
Barley(6)	*Hordeum vulgare*	6	1.4	2.4	80
Soybean	*Glycine max*	3	2.1	5.6	73
Soybean	*Glycine max*	3	2.1	5.0	
Pea(5)	*Pisum sativum*	5	-	3.9	78
Pinto bean	*Phaseolus vulgarus*	5	-	2.2	
Navy bean	*Phaseolus vulgarus*	5	-	3.2	
Alfalfa	*Medicago sativa*	3.5	6.3	17.9	75
Lentil	*Lens esculenta*	3.5	3.2	3.3	
Mung bean	*Phaseolus aureus*	3.5	5.2	12.8	
Soybean	*Glycine max*	3.5	2.5	6.0	
Alfalfa(light)	*Medicago sativa*	5	0.7	2.3	39
Alfalfa(dark)	*Medicago sativa*	5	0.7	2.3	
Yellow pea	*Pisum sativum*	4	0.2	0.5	37
Lentil	*Lens esculenta*	4	0.3	1.4	
Faba bean	*Vicia faba*	4	1.0	1.3	
Mung bean(light)	*Phaseolus aureus*	2.5	1.8	2.3	48
Mung bean(dark)	*Phaseolus aureus*	2.5	1.8	2.3	

[a] Water cultured. [b] Sand cultured. [c] μg/seedling(d.b.).

Number of varieties or seed lots in parantheses.

that the bound form required hydrolysis before extraction. Swaminathan[89] was among the first to quantitate niacin in wheat, barley, and rice, all of which contained about 45 to 50 μg/g (dry basis). Edible legumes contained 13 to 26 μg/g, with peas containing the least and chickpeas the most. White flour, maize, oatmeal, millet and raw milled rice contained about 1/2 to 1/3 as much of the vitamin as did most legumes.

Niacin in dried brewers yeast and food-yeast (Torula utilis) varied from 430 to 620 and 200 to 340 μg/g, respectively (Table 5). In addition to hydrolyzing the bound form of niacin, the germination process invariably promotes increases in total niacin (Table 5).

Although dried yeast is among the most potent sources of niacin with about 200 to 600 μg/g niacin, cereals and legumes are good sources, ranging from 20 to 60 or more μg/g (Table 5). Generally, cereals contain about twice as much niacin as legumes. The peanut is an exception, and often contains as much as 125 to 175 μg/g of niacin, about 3 times as much as most cereals.

Klatzkin et al[90] emphasized that cereals vary widely in their nicotinic acid content due both to genetic (varietal) differences and to growing conditions. Maize and oats were singled out as poor sources of niacin, rice and barley better and whole wheat as the best source among staple cereals. Twelve days of relatively slow germination resulted in 58 to 108% increases in oats, 0 to 52% increases in rice, while wheat, barley and maize remained unchanged in niacin content.

Kylen and McCready[75] assayed the nicotinic acid content of seeds and sprouts of alfalfa, lentils, mung beans and soybeans. It was lowest in alfalfa but increased more in that species than in any other seed studied, from about 19 to 137 μg/g. Prudenta and Mabesa[48] found that light did not alter the rate of niacin synthesis in germinating mung beans. Niacin increased from about 6.2 to 8.5 μg/g after 3-1/2 days of germination.

Germination from 2 to 5 days invariably enhances total niacin content of edible cereals and legumes. The average niacin content of seed in Table 5 is 27 μg/g. Germination increased that value to 53 μg/g.

Biotin

Table 6 summarizes the biotin content of some edible seeds and sprouts. Generally legumes contain about 0.60 μg biotin per g, ranging from about 0.20 to 1.3 μg/g. Oats contain the most biotin among the cereals, an average of 0.90 μg/g. Generally cereals contain about 30% as much biotin as legumes.

Two to five days of germination doubles the biotin content of edible cereals as well as legumes. The average biotin content of ungerminated seed is about 0.48 μg/g, and of germinated seed, about 1.00 μg/g (dry basis).

Pyridoxine

Swaminathan[93] found dried brewers yeast to be the richest source of vitamin B_6, 54 μg vitamin B_6 per g yeast. Rice polishings and fresh sheep liver were also good sources, containing 20 μg and 14 μg of vitamin B_6 per g polishings and liver, respectively. Whole cereals and pulses contained about 7 to 10 μg of B_6 per g. More than 1/2 of the vitamin was lost in milled and refined samples of wheat and rice. Among legumes, chickpeas contained the most vitamin B_6 (10.6 μg/g), while soybeans contained the least (8.6 μg/g). In addition Sarma[94] indicated wheat and sorghum contained about 8 μg/g of vitamin B_6, rice about 4 μg/g, while bengal gram, green gram and black gram each contained about 10 μg/g. Brewers yeast contained about 50 μg/g, while soybeans contained about 8 μg/g. Sarma emphasized that the microbiological method compared favorably with the best chemical methods of the day. Burkholder and McVeigh[65] reported their results for pyridoxine were not entirely satisfactory because of interfering substances in the samples, but in preliminary work found large increases during early germination of wheat, barley, corn, oats, soybeans, lima beans, greeneye peas, mung beans and peas. Others have reported significant increases in pyridoxine after 2 to 6 days germination (Table 7).

Pantothenic Acid

Davis et al[76] reported that pantothenic acid remained unchanged when dry weight losses of germinated wheat and barley were considered. Other researchers studied the effect of germination on pantothenic acid content of edible cereals and legumes and found it increased in come species and remained constant in others (Table 8).

Folic Acid

Cheldelin and Lane[70] reported losses of 46 and 17% in folic acid in blackeye peas and lima beans, respectively, after 48 h germination. The concentration of folic acid in cottonseed increased by 110% after 36 h germination. Burkholder[71] reported increases of folic acid in dormant

Table 5. Nicotinic acid content of some edible seeds and sprouts.

Common Name	Scientific Name	Germ. Time (Days)	Nicotinic Acid Seed (μg/g)	Nicotinic Acid Sprout (μg/g)	Reference Number
Soybean(immature)	Glycine max	0	40	-	66
Soybean(mature)	Glycine max	0	20	-	
Lima bean(3)	Phaseolus lunalatus	0	22	-	67
Blackeye pea	Vigna cylindrica	0	28	-	
Pea(2)	Pisum sativum	0	34	-	
Peanut(2)	Arachis hypogea	0	147	-	
Lentil	Lens esculanta	0	25	-	
Wheat	Triticum vulgare	5	83	72	69
Corn	Zea mays	5	20	30	
Soybean	Glycine max	5	26	37	
Mung bean	Phaseolus aureus	5	30	51	
Barley	Hordeum vulgare	5	72	129	
Corn	Zea mays	5	17	40	
Oat	Avena byzantina	5	11	48	
Soybean	Glycine max	5	27	49	
Lima bean(2)	Phaseolus lunalatus	5	13	35	
Greeneye pea	Vigna cylindrica	5	20	60	
Mung bean	Phaseolus aureus	5	26	70	
Pea	Pisum sativum	5	31	32	
Wheat[a]	Triticum vulgare	6	62	121	
Corn[a]	Zea mays	12	15	50	
Wheat[b]	Triticum vulgare	5	60	104	
Corn[b]	Zea mays	5	12	38	
Oat[b]	Avena byzantina	5	4	40	
Barley[b]	Hordeum vulgare	5	62	102	
Blackeye pea	Vigna cylindrica	2	14	51	70
Lima bean	Phaseolus lunalatus	2	11	26	
Cottonseed	Gossypium ?	2	16	24	
Oat	Avena byzantina	5-6	8	44	71
Wheat	Triticum vulgare	5-6	62	103	
Barley	Hordeum vulgare	5-6	68	115	
Corn	Zea mays	5-6	10	40	
Barley(7)	Hordeum vulgare	6	86	94	76
Wheat(5)	Triticum vulgare	6	59	67	
Pea(3)	Pisum sativum	3	28	45	13
Soybean(6)	Glycine max	3	18	33	
Oat	Avena byzantina	3	0.16[c]	0.9[c]	77
Soybean(7)	Glycine max	4	21	31	72
Mung bean	Phaseolus aureus	4	23	67	

Table 5 continued.

Common Name	Scientific Name	Germ. Time (Days)	Nicotinic Acid Seed (μg/g)	Nicotinic Acid Sprout (μg/g)	Reference Number
Soybean	*Glycine max*	3	27	46	73
Soybean	*Glycine max*	3	21	51	
Ghasi mung(3)	*Phaseolus radiatus*	2,4	24	64,66	91
Mash Kalai	*Phaseolus mungo*	2,4	15	24,32	
Motor(3)	*Pisum sativum*	2,4	12	35,44	
Chola(3)	*Cicer arietinum*	2,4	10	21,26	
Khesarii	*Lathyrus sativus*	2,4	4	18,26	
Mushuri	*Lens esculenta*	2,4	17	28,34	
Barbati	*Vigna sinensis*	2,4	18	38,41	
Soyabean	*Glycine soya*	2,4	18	49,54	
French bean	*Dolichos Sp-red*	2,4	8	20,25	
French bean	*Dolichos Sp-white*	2,4	9	36,42	
Chim	*Dolichos lablab*	2,4	0	26,31	
Arahar	*Cajanus indicus*	2,4	7	23,26	
Joar	*Sorghum vulgare*	2,4	6	13,16	
Khesarri	*Lathyrus sativus*	4	33	49	92
Kalai	*Phaseolus mungo*	4	21	45	
Kablimatar	*Pisum sativum*	4	31	48	
Arahar	*Cajanus indicus*	4	28	47	
Mung	*Phaseolus radiatus*	4	21	66	
Chola	*Cicer arietinum*	4	17	25	
Musur	*Lens esculenta*	4	16	31	
Pea(5)	*Pisum sativum*	5	-	24	78
Pinto bean	*Phaseolus vulgarus*	5	-	23	
Navy bean	*Phaseolus vulgarus*	5	-	18	
Alfalfa	*Medicago sativa*	3.5	19	137	75
Lentil	*Lens esculenta*	3.5	35	40	
Mung bean	*Phaseolus aureus*	3.5	20	78	
Soybean	*Glycine max*	3.5	33	41	

[a] Water cultured. [b] Sand cultured. [c] μg/seedling.

corn, barley, oats and wheat after 5 to 6 days germination of 350, 245, 550, and 278%, respectively. Banerjee et al[93] reported that 4 days of germination materially reduced the folic acid content of 7 pulses. Black gram and mung beans contained the most folic acid before (144 and 145 mg/100 g) and after (114 and 93 mg/100 g) germination. Chickpeas,

Table 6. Biotin content of some edible seeds and sprouts.

Common Name	Scientific Name	Germ. Time (Days)	Biotin Seed (μg/g)	Biotin Sprout (μg/g)	Reference Number
Barley	*Hordeum vulgare*	5	0.40	1.20	69
Corn	*Zea mays*	5	0.30	0.70	
Oat	*Avena byzantina*	5	1.20	1.80	
Soybean	*Glycine max*	5	1.10	3.50	
Lima bean(2)	*Phaseolus lunalatus*	5	0.10	-	
Greeneye pea	*Vigna sinensis*	5	0.40	1.10	
Mung bean	*Phaseolus aureus*	5	0.20	1.00	
Pea	*Pisum sativum*	5	-	0.50	
Wheat[a]	*Triticum vulgare*	12	0.32	1.20	
Corn[a]	*Zea mays*	12	0.28	1.50	
Wheat[b]	*Triticum vulgare*	5	0.18	0.30	
Corn[b]	*Zea mays*	5	0.10	0.38	
Oat[b]	*Avena byzantina*	5	0.60	0.98	
Barley[b]	*Hordeum vulgare*	5	0.21	0.60	
Soybean(immature)(6)	*Glycine max*	0	0.54	-	66
Soybean(mature)(6)	*Glycine max*	0	0.61	-	
Blackeye pea	*Vigna cylindrica*	2	0.22	0.41	70
Lima bean	*Phaseolus lunalatus*	2	0.11	0.20	
Cottonseed	*Gossypium* ?	2	0.29	0.28	
Oat	*Avena byzantina*	5-6	0.90	1.40	71
Wheat	*Triticum vulgare*	5-6	0.17	0.36	
Barley	*Hordeum vulgare*	5-6	0.31	0.91	
Corn	*Zea mays*	5-6	0.21	0.54	
Soybean(7)	*Glycine max*	4	0.80	1.50	72
Mung bean	*Phaseolus aureus*	4	0.20	0.60	
Khesari	*Lathyrus sativus*	4	0.80	0.70	92
Kalai	*Phaseolus mungo*	4	0.80	1.00	
Kablimatar	*Pisum sativum*	4	0.80	0.92	
Arhar	*Cajunus indicus*	4	0.80	0.96	
Mung	*Phaseolus radiatus*	4	0.80	1.60	
Chola	*Cicer arietinum*	4	1.00	1.30	
Musur	*Lens esculenta*	4	1.30	1.50	

[a] Water cultured. [b] Sand cultured.

Table 7. Pyridoxine content of some edible seeds and sprouts.

Common Name	Scientific Name	Germ. Time (Days)	Pyridoxine Seed (μg/g)	Pyridoxine Sprout (μg/g)	Reference Number
Soybean(immature)	*Glycine max*	0	3.5	-	66
Soybean(mature)	*Glycine max*	0	6.4	-	
Blackeye pea	*Vigna cylindrica*	2	2.0	4.6	70
Lima bean	*Phaseolus lunalatus*	2	6.0	9.5	
Cottonseed	*Gossypium* ?	2	0.9	2.6	
Oat	*Avena byzantina*	5-6	0.3	1.8	71
Wheat	*Triticum vulgare*	5-6	2.6	4.6	
Barley	*Hordeum vulgare*	5-6	0.2	0.5	
Corn	*Zea mays*	5-6	0.7	0.8	
Soybean(7)	*Glycine max*	4	11.8	15.6	72
Mung bean	*Phaseolus aureus*	4	11.9	14.4	

which contained 125 mg folic acid per 100 g before germination, contained only 9 mg per 100 g after 4 days sprouting, much less than any of the other legumes.

Legumes generally contain more pyridoxine (vitamin B_6) than do the cereals; although wheat generally contains about as much as the legumes. Germination for 2 to 5 days generally enhances vitamin B_6 content in the edible seeds by 50 to 100%.

Legumes contain 2 to 3 times as much pantothenic acid as do cereals. Germination increases the pantothenic acid content in cereals, but not in most legumes. As with pyridoxine, more research is required to adequately evaluate the effect of germination on pantothenic acid in edible cereals and legumes. The folic acid data indicates that germination generally reduces that vitamin.

Vitamin B_{12}

When Esh[95] fed lentils at 18% protein level to rats, he found supplementary amounts of vitamin B_{12} significantly increased growth and biological value of the protein. In

Table 8. Pantothenic acid content of some edible seeds and sprouts.

Common Name	Scientific Name	Germ. Time (Days)	Pantothenic Acid Seed (µg/g)	Pantothenic Acid Sprout (µg/g)	Reference Number
Soybean(immature)	Glycine max	0	12	-	66
Soybean(mature)	Glycine max	0	12	-	
Blackeye pea	Vigna cylindrica	2	11	17	70
Lima bean	Phaseolus lunalatus	2	9	20	
Cottonseed	Gossypium ?	2	11	22	
Oat	Avena byzantina	5-6	8	22	71
Wheat	Triticum vulgare	5-6	8	13	
Barley	Hordeum vulgare	5-6	5	10	
Corn	Zea mays	5-6	4	8	
Soybean(7)	Glycine max	4	36	34	72
Mung bean	Phaseolus aureus	4	16	34	
Khesari	Lathyrus sativus	4	26	32	92
Kalai	Phaseolus mungo	4	35	43	
Kablimatar	Pisum sativum	4	21	19	
Arhar	Cajunus indicus	4	15	23	
Mung	Phaseolus radiatus	4	25	24	
Chola	Cicer arietinum	4	13	15	
Musur	Lens esculenta	4	16	14	

addition, vitamin B_{12} increased the protein efficiency ratio (PER) value of the lentil hydrolysate even after it had been supplemented with methionine (0.17%), tryptophan (0.05%), and threonine (0.075%). PER increased from 2.06 to 2.60 (Table 9).

Rohatgi et al[96] reported vitamin B_{12} values ranged from 0.36 to 0.61 µg per 100 g in 7 Indian pulses. The vitamin B_{12} concentration in lentils, peas and chickpeas increased 451, 556, and 443% after 4, 4, and 2 days germination, respectively. In Cajanus indicus, it maximized after 2 days with an increase of only 28%, however in three other legumes, Lathyrus sativus, mung beans, and black gram, it increased 132, 151 and 186%, respectively at 4 days germination.

Table 9. Influence of B_{12} on the protein regeneration when an enzymic digest of lentil protein was fed to protein depleted adult rats for 12 days. (Average of 6 rats in each group).[95]

Digest and supplement*	Intake of 5% Protein Solution %	Total Protein Intake g	Weight Recovered g	Protein Efficiency Ratio
Lentil Hydrolysate	80	8.5	Negative	-
" + Met	89	10.2	16.6	1.63
" + Met + Try	90	10.6	21.4	1.90
" + Met + Try + Thr	88	10.17	21.0	2.06
" + Met + Try + Thr + B_{12}	92	10.78	28.0	2.60

* The lentil hydrolysate solution containing 5% protein was supplemented with 0.17% DL methionine, .05% L tryptophan, 0.75% L. threonine and 8 microgram percent vitamin B_{12} in respective groups.

Vitamin B_{12} values reported by Rohatgi et al[96] were questioned by Ramachandran and Phansalkar[97] after extracting the four pulses using the procedure of Rohatgi et al[96] and subjecting the pulses to treatment with 0.2 N NaOH at 100°C for 30 min. They concluded that B_{12} activity was not diminished by the alkali treatment, and that the activity was not cyanocobalamin but some other purine or desoxyriboside which could effectively replace B_{12} for _L. leichmannii_, the microorganism used by Rohatgi et al[96] to assay for vitamin B_{12}.

Lavate and Sreenivasan[98] analyzed mung beans after 0, 2, and 4 days germination and found vitamin B_{12} activity in both the dormant and germinated seeds to be alkali-stable desoxyribosides, and therefore not true vitamin B_{12}. In 1979, Hofsten[99] discussed legume sprouts as a source of protein and other nutrients. He stated that these sprouts are often a "remarkably good source of vitamin B_{12}." In that article it was not stated which legume sprouts are remarkably good sources of B_{12} nor was it clear who completed the B_{12} assays. Hofsten[100] also mentioned that he or his associates had found B_{12} in unnamed sprouts. More research is required to substantiate or refute studies which affirm the presence of B_{12} in seeds and sprouts.

FAT SOLUBLE VITAMINS

The four fat soluble vitamins, A, D, E and K are isoprenoid compounds, about which relatively little is known concerning their exact biological functions. No specific coenzyme function for any of the fat soluble vitamins has yet been established.[101]

Vitamin A (Retinol)

Vitamin A activity in mammals is supplied by retinols and also α-, β-, and γ-carotene. The carotenes themselves have no intrinsic vitamin A activity but may be enzymatically converted into the active form of the vitamin in the liver or intestine. No articles were reviewed which directly related the effects of germination of cereals or pulses to their vitamin A activity; however the effects of germination on seed carotene content have been studied.

Carotene

Miller and Hair[15] reported that mung bean sprouts contained about 110 to 120 carotene units per lb, compared to 750 to 3000 units in lettuce and 2500 to 3500 units in fresh green peas. Lee and Whitcombe[79] reported 4 immature and germinated mature (raw and cooked) soybean varieties varied in carotene content from about 3.53 to 1.19 μg/g (wet basis). Wai et al[73] reported soybeans increased in carotene content from 1.2 to 4.3 μg/g and 1.3 to 2.0 μg/g during two different 3 day germination regimes.

Chattopadhyay and Banerjee[102] found 4 days germination of legumes and 7 days germination of cereals had an equilibrating effect on total carotene content. For example among the legumes, 4 chickpea samples, which contained 1.95, 2.05, 2.95 or 3.10 mg/100 g, increased to 3.00, 4.30, 4.20, and 4.30 mg/100 g, respectively, after germination. That equilibrating effect due to germination was more pronounced among cereals assayed. Rice, wheat and corn contained 0.35, 0.45 and 4.00 mg carotene/100 g; however, after 7 days germination they contained 3.95, 4.05 and 6.15 mg/100 g, respectively.

Fordham et al[34] reported the carotene content of a number of pea and bean varieties varied from 3.16 to 37.35 μg/100 g, and 0.23 to 16.34 μg/100 g, respectively. Only the 4 to 6 day hypocotyls were assayed for carotene and those values varied from 2.43 to 45.36 μg/100 g for peas and from 3.64 to 59.47 μg/100 g for beans. Among the pea cultivars, Early Alaska possessed the most carotene (37.35 μg/100 g) but the 4 to 6 day old hypocotyl contained only 4.35 μg/100 g (dry basis). Another cultivar, Mammoth Melting, contained 13 μg/100 g in the seed and 45.36 μg/100 g in the hypocotyl fraction. The same lack of direct correlation existed between the carotene content of beans and their hypocotyls.

Vitamin D

No reports were found on the effects of germination on the vitamin D content of cereals or legumes.

Vitamin E (Tocopherol)

Chattopadhyay and Banerjee[103] reported increases of about 25% in tocopherol content of 7 pulses and 3 cereals after 4 days germination. Although dry weight losses were not reported, some apparent increases in vitamin E may be accounted for by dry weight losses from leaching and respiration during early germination.

Vitamin K

Dam et al[104] reported that peas contained about 14 vitamin K units/g which increased to 335 units after 16 days growth in light and to 95 vitamin units/g when grown in the dark. Erkama and Pettersson[105] indicated vitamin K in etiolated peas increased by a factor of about 25 in light after germinating 17 days and increased by a factor of about 12 to 13 in the dark. Banerjee et al[106] found 6 species of legumes including 2 samples of chickpea increased 13 to 84% in vitamin K content after 4 days germination, with absolute values of about 0.30 mg/100 g. Chickpea samples ranged from 0.30 to 0.50 mg/100 g after germination.

OTHER WATER-SOLUBLE GROWTH SUBSTANCES

In addition to the B vitamins, whose coenzyme functions are reasonably well known, there are several other water-soluble substances which may be growth factors, and are ubiquitous in plants and their seeds. Among those factors are inositol and choline.

Inositol: Phytate and Phytase Interaction

In plant seeds wherever phytic acid is present phytase is also found. However, germination is required to activate phytase. In cereal and legume grains a considerable proportion of the total P is present as phytate. In many food grains 65 to 80% of the total P occurs as phytin P. The relatively high content of phytate in cereals and legumes is important in human and animal nutrition since phytate has long been known to interfere with the metabolism and absorption of certain minerals, notably Fe, Zn, and Ca.

A principal physiological function of phytase is to provide inorganic phosphate (P_i) from phytate during the initial stages of germination. Peers[107] concluded that the proportion of P_i to phytate increased progressively when barley, wheat, oats, maize and rice were germinated.

Burkholder[66] indicated that the inositol content of oats and corn approximately doubled after 5 to 6 days germination. In wheat, inositol increased about 15 to 20% while that in barley remained nearly constant, if dry weight losses are considered. Cheldelin and Lane[69] reported that the inositol content of cottonseed decreased after 48 h germination but increased in blackeye peas and lima beans by 77 and 110%, respectively. Burkholder and McVeigh[72] assayed mung beans and 6 soybean cultivars for inositol content. After 4 days of germination, mung bean inositol increased from 1850 to 3649 µg/g; while it increased in the soybean cultivars from 2290 to 3356 µg/g.

During the past 30 years much research on the relationship of germination of food grains to phytate hydrolysis and mineral bioavailability has appeared. Belavady and Banerjee[108] discovered 7 species of legumes germinated for 5 days doubled in phosphatase activity. Phytin P was diminished by approximately 50 to 60% in all 7 species. Total P among the 7 legumes ranged from 300 to 400 mg/100 g, with 15 to 35% of the total P bound in phytate.

Banerjee et al[106] found that the inositol content in 7 legumes species increased after 2 and 4 days germination. In ungerminated seeds mung beans contained the least (70 mg/100 g) while chickpeas contained the most inositol (240 mg/100 g). After germination chickpeas contained 330 mg/100 g while lentils, *Lathyrus sativus*, black gram, peas, red gram and mung beans contained 210, 200, 170, 170, 150, and 110 mg/100 g, respectively.

Darbre and Norris[109,110] recalculated actual inositol gains obtained by previous workers to correct for dry weight losses of seeds during germination. For example, they recalculated Burkholder's[66] inositol values, suggesting that the real increases were more nearly 12.2 to 19.1 μg of inositol/seed. Referring to their own studies, those authors noted total inositol losses were 10 to 20% after 10 days germination, with gains in the free inositol form but with greater losses in the bound form of the growth substance. It is ironic in the context of their criticism of earlier workers that their own dry weight losses amounted to only 8.7% after 10 days germination, which indicated that their germination conditions were relatively poor or that a relatively high percentage of their seeds were sickly or lacked viability.

Studying lettuce seeds, Mayer[111] reported that phytate was rapidly exhausted after just 72 h germination. Phytase activity also maximized just after 72 h of sprouting, with optimum activity at pH 5.2. Ergle and Guinn[112] noted a sizable reduction in the phytate and P_i content in cottonseed cotyledon and hypocotyl fractions after 4 and 6 days germination. Sobolev[113] reported a considerable portion of phytin was decomposed during 5 days of wheat and sunflower germination. Hydrolysis of phytin during germination is accomplished apparently "by two phosphatases, one of which (phytase) is responsible for the initial stage of phytin decomposition to the stage of the formation of ITP and IDP and the other accomplishes a further complete dephosphorylation of inositol phosphates." Gibbons and Norris[114] reported that the total inositol content of whole beans showed a marked decrease during the first 2 weeks of growth, a phenomenon which had been observed during the germination of cabbage, maize, mung beans, peas, soybeans, wheat when grown in the light, etiolated pea and mung bean seedlings,[115] and in oats.[110] Gibbons and Norris[114] found an increase in total inositol in roots and leaves "but whereas in the leaves this rise is due to an influx of free inositol, that in the roots is due to bound forms." That finding may reflect functional differences (in the physiological sense) of bound and free forms of the compound. Inositol hexaphosphate has an affinity for Ca^{2+}, Mg^{2+}, and K^+ ions in the phytin complex. The root system is involved in the ion uptake of the plant, and the presence of relatively large quantities of inositol phosphates in the roots may be exerting an influence on the ionic balance.

Milailovic et al[116] found the esters of _myo_-inositol phosphates, from mono- to penta-, in wheat seedlings. Those esters are produced _in vitro_ when sodium phytate and phytase are incubated in solution.[117] Shrivastava[118] reported phytase activity increased 290% with gibberellic acid (GA) and 220% without GA during 96 h germination of barley (after steeping 72 h, 15°C). Matheson and Strother[119] found that wheat phytase activity increased without GA by 320% after 3 days germination (25°C), and that it was depleted after 14 days sprouting. Bianchetti and Sartirana[120] showed phytase activity of wheat germ and scutellum maximized without GA after 30 h germination (2 h steep, 20°C, then 28°C in dark with 0.5% sucrose).

Mandal and Biswas[121] found that phytase activity in germinating mung beans maximized at about 72 h and increased slowly in the embryo fraction. They reported that the synthesis of phytase in the cotyledon was dependent on the synthesis of new RNA in the cotyledon fraction. Chen and Pan[122] reported the phytate content of soybeans, Early Alaska peas, and Dwarf Gray peas was 2.48, 1.86 and 1.13% of total seeds. Germination for 5 days reduced the phytate content of those legumes to 1.93, 1.20 and 0.59% of the seeds. The phytase activity of soybeans was higher in resting seed than in either pea species. Five days of germination increased the phytase activity in Early Alaska peas by about twice as much as in the Dwarf Gray peas. The phytase activity of the Early Alaska peas nearly equalled the activity in the soybeans, which doubled during germination.

Reddy et al[123] assayed black gram for phosphorus and reported it was about 0.5% of the total seed weight, with about 80% in the form of phytate P. About 50% of the phytate was hydrolyzed by the tenth day of germination. Boiling for 45 min did not change the phytate to P ratio. However there was a loss of P, as well as Ca, Mg, and Fe, due to leaching during initial hydration with distilled water. The first 5 min of boiling also resulted in a major loss of minerals.

Ganesh Kumar et al[124] reported that chickpeas contained about 2/3 as much P as green gram and cowpeas, and about 1/3 as much phytic acid P as green gram and 1/2 as much as cowpeas. The phytin P to total P ratio in green gram, cowpeas and chickpeas after 72 h germination decreased about 25, 78, and 23%, respectively.

Effect of Heat on Phytate Hydrolysis

That an "incubation" period would result in reducing phytate in some edible seeds was first suggested by Peers.[107] Mandal et al[125] studied the effects of incubating mung beans at different temperatures for 1 h. Increasing amounts of P were released from phytate with temperature increase to 57°C, but decreased rapidly above 57°C. At 65°C the phytase activity was about 1/3 that at 57°C.

Kon et al[126] explored ways to initiate and maximize hydrolysis of phytate as well as some oligosaccharides in small white beans by varying the pH of the medium, by incubating at 55°C for 20 h, by boiling 2 h, by adding 1.2% wheat germ phytase, or by some combinations of those treatments. Kon et al[126] successfully hydrolyzed phytic acid P to P_i. In two experiments they hydrolyzed 77 and 93% of the phytate and in 4 other experiments 60 to 70% of the phytate was broken down to P_i. Chang et al[127] noted that the same species of small white beans contained about 1% phytate, of which 70% was reportedly water soluble. They incubated pre-soaked beans for 10 h in water at 60°C and reduced phytate by about 75%. Incubation in water-saturated air overnight was reported to reduce phytate by 50%. A similar reduction in phytate was observed when a temperature of 60°C was used to incubate mung beans, lima beans, and wheat; although only a 37% reduction in soybean phytate was observed.

Tabekhia and Luh[128] also studied the effects of germination and variable cooking conditions on phytate retention in blackeye, red kidney, mung, and pink beans. The soaked beans (12 h, 24°C) hydrolyzed phytate only slightly. Seventy-two h germination reduced the phytin P of those beans by 37, 34, 30, and 32%, respectively. Germination for 120 h further hydrolyzed phytate in blackeye and pink beans (87 and 47%, respectively) but did not change the phytate content in red kidney or mung beans. When cooked for 3 h at 100°C, the phytate content of the 4 types of dry beans was only moderately effected but cooking for 3 h at 115°C in sealed cans very materially reduced phytate by 92, 70, 68, and 75%, respectively.

Finney et al[129] observed that 12 genetically highly divergent wheat variety composites contained phytate ranging from 0.58 to 1.0% (dry basis), with red wheats containing the most. Four white wheat composites were among the five with lowest phytate content. Five days of germination lowered the phytate content by 40 to 60%. Non-phytate P in the 12 wheat varieties varied from 0.012 to 0.026% of total P. Five days of germination increased the non-phytate P to between 0.046 and 0.106%.

Choline

Chattopadhyay and Banerjee[130] established the choline content of seven legume species and three cereals. Legumes generally contained 200 to 245 mg choline/100 g with the exception of black gram and *Phaseolus roxburgi* which contained about 100 mg/100 g. Most of the 15 to 25% increases reported by the authors is atributable to dry weight losses due to respiration and leaching during hydration and germination.

MINERALS

Cravioto et al[131] reported Ca and Fe contents of numerous Mexican foods, including corn, rice, peanuts, wheat, chick-peas, and sesame seeds, which contained 7, 8, 49, 50, 100 and 417 mg Ca/100 g, and 2.8, 0.8, 2.1, 9.2, 9.3, and 8.4 mg/100 g Fe, respectively. In an excellent review of legumes in human nutrition, Patwardham[132] stated that "legumes are comparatively poor sources of dietary calcium, the highest value reported being less than 300 mg/100 g, more often the calcium content is in the neighborhood of 100 mg/100 g or less." However legumes are reasonably good sources of dietary Fe, since they contain between 2 and 10 mg/100 g. Wheat grown in the United States contains about 50 mg/100 g Ca, and about 10 mg/100 g Fe, with varietal and environmental variations producing a range from 30 to 80 mg/100 g Ca and from 3 to 16 mg/100 g Fe in 279 wheat samples from throughout the U.S. grown in 1973, 1974, or 1975.[133] Thus whole wheat is a poorer source of Ca than most legumes, but a somewhat better source of dietary Fe.

Gapalan et al[134] listed the Ca content in several Indian foodstuffs, including sesame seed (*Sesamum indicum*), curry leaves (*Murraya koe nigii*), drumstick leaves (*Moringa oleifera*), amaranth (*Amarantus tricolor*), fenugreek (*Trigonella foenumgraecum*), and ragi (*Eleusine coracana*) which contained 1450, 830, 440, 397, 360 and 344 mg/100 g, respectively. Fordham et al[34] analyzed a number of common pea and bean species and varieties and found them to be fair sources of Fe with all but two samples containing from 4.4 to 9.5 mg/100 g. Ca values for peas ranged from 68 to 118 mg/ 100 g (average 93 mg), and for beans from 88 to 318 mg/100 g (average 166 mg). Mg, P, and K contents of the peas and beans were approximately in a ratio of 1:2:4, with beans containing about 2 to 40% more of each element. Average values for Mg, P and K in peas were 141, 366, and 861, and for beans 186, 403, and 1175 mg/100 g, respectively. While Fordham et al[34] reported that soybeans contained more Ca and Fe (318 and 9.5 mg/100 g, respectively) than other samples tested, Kylen and McCready[75] found soybeans contained more Ca but less Fe (280 and 1.6 mg/100 g, respectively) than the other samples they analyzed.

Mn ranged from 7.0 to 20.8 mg/100 g for peas and averaged about 15.5 mg/100 g; while for beans they ranged from 12.5 to 34.3 mg/100 g and averaged about 21.1 mg/100 g.[34] Mg content of about 300 whole wheat samples ranged from 90 to 210 mg/ 100 g, and averaged about 150 mg/100 g.[133] P ranged from 350 to 910 mg/100 g, and averaged about 580 mg/100 g; while K ranged from 320 to 610 mg/100 g, and averaged about 430 mg/ 100 g. Thus whole wheat generally contains more Mg than peas and less than beans. Wheat also contains more P than either peas or beans, but only about 50% as much K as peas and about 37% as much K as beans.

In the ungerminated legumes, chickpeas contained more than twice the Ca as green gram and cowpeas (about 5.5 versus 2.4 mg/100 g) and about the same amount of Mg as green gram and cowpeas (about 7 to 8 mg/100 g).[124] Von Ohlen[287] noted that K moved out of the cotyledons of germinating soybean more rapidly than P and Mg. Fordham et al,[34] analyzing only the hypocotyl portion of the sprouted seeds, noted that among all the peas and beans tested, mung bean sprouts contained 3 times more Fe than other samples. More variability occurred in the Mn concentration of the sprouted portion of peas and beans. While there was a 3-fold variability among dry peas or beans, there was a 14-fold variability among pea sprouts and a 7- to 8-fold variability among bean sprouts. Whole wheat is a relatively good source of dietary Zn, ranging from about 2 to 7.8 mg/100 g, with an average of about 4.5 mg/ 100 g.[133] Kylen and McCready[75] reported Zn values of 7.53,

5.08, 4.20 and 6.76 for alfalfa, lentils, mung beans and soybeans, about the same amount as whole wheat.

Germination and Mineral Bioavailability

Singh and Banerjee[135] reported that germination increased the protein-free Fe variably for 8 Indian legumes. After 72 h of germination cowpeas contained the highest (74%) percentage of protein-free Fe and chickpeas the least (13%). Ionizable Fe increased with both sprouting and fermentation.[136] Germination increased the ionizable Fe of green gram, peas, pink beans, chickpeas and cowpeas by 17.5, 23.6, 20.3, 36.6, and 7.4%, respectively.

Beal et al[137] evaluated the effects of germinating peas on Zn bioavailability in rats. Dehydrated blanched peas and freeze-dried 8-day germinated peas with phytate/Zn molar ratios of 35 and 23, respectively, were incorporated into albumin-based diets to obtain marginal Zn levels of 10 ppm. The diets and a non-phytate control at 10 ppm Zn were fed at two levels of Ca for 28 days. At the adequate level of Ca (0.75%) weight gain of all rat groups did not differ significantly. However at 1.50% Ca germinated peas and the non-phytate control promoted equal weight gains; while weight gain from blanched, ungerminated peas was 50% less. For both levels of Ca, Zn bioavailability in rat tibia increased when rats were fed germinated peas.

Effects of Germination on Mineral Content

Depending on methods of steeping and germination, some minerals may be leached from or absorbed by hydrating and germinating seeds. In an investigation of Chinese foods, Embrey[62] parenthetically called attention to a large increase in Ca that occurred during germination of mung beans and soybeans. The beans, the author pointed out, were sprouted in the hard city water of Peking which contained an abundance of Ca and Mg salts. When considering the importance of Ca in physiological functions and the deficiency of Ca in most cereals and legumes, that information could be of special value in human and animal nutrition. Fordham et al[34] reported that the increased concentration of Ca in sprouted seeds of two soybean varieties was directly related to the concentration of chlorinated lime in the steep water. Beal et al[137] noted an increase in Ca from 400 to 1640 ppm when peas were automatically rinsed every two hours over an 8 day period of germination with tap water known to contain an intermediate level of Ca salts. Kylen et al[75] reported increases in Ca for alfalfa germinated 3 days (from 147 to 239 mg/100 g). Lentils, mung beans, and soybeans did not change in Ca concentration under those conditions.

Often minerals are lost to the steep medium, especially if distilled water is used to inbibe seeds. Losses of 40% of the Ca and Mg from green gram and cowpea and 20% of those minerals from chickpeas, mostly due to leaching during steeping occurred during 72 h of germination.[124]

CARBOHYDRATES

Carbohydrate Metabolism

Appearance of reducing sugars and an increase of starch in the hypocotyl, root cap and cotyledon were among the first changes in carbohydrates noted during germination of soybeans.[287] As early as 1943,[138] evidence was obtained for the operation of two distinct respiratory mechanisms which controlled the growth and respiration of oat seedlings. That observation was based on the insensitivity of the seedlings to potassium iodoacetate, a sugar poison, and absence of stimulation by pyruvate and C_4 dicarboxylic acids. During the early stage of germination oxygen uptake as well as growth was inhibited by sodium azide, but after 72 h of sprouting the inhibitory effect was lost. At that point, sugar is utilized, the seedling becomes sensitive to iodoacetate poisoning, sugar intermediates such as pyruvate and the dicarboxylic acids stimulate growth, lipid utilization diminishes, and reducing sugars are utilized rapidly.

In germinated castor bean, all enzymes necessary to bring about conversion of fat to carbohydrate: Fat → fatty acyl CoA → acetyl CoA → [glyoxylate cycle] → malate phosphoenolpyruvate → [glycolysis] → carbohydrate are present.[139] Cavin and Beevers[140] concluded that the "tricarboxylic acid cycle is not operative in [castor bean] and showed that the major fate of acetate is conversion to sucrose by way of dicarboxylic acids produced in the glyoxylate cycle."

Young and Varner[141] reported that phosphatase reached maximum activity 5 days after germination in Alaska peas but amylase was not synthesized until 2 to 3 days after germination, increasing exponentially to the completion of the 8 day study.

Sprague and Yemm[142] reported that oxygen uptake in germinating peas rapidly increased for 2 h, then decreased so that after 48 h very little oxygen was being utilized. During the first 10 to 12 h CO_2 production increased steadily. It was estimated that 2/3 of that CO_2 arose from alcoholic fermentation. Once the testa split aerobic conditions appeared to prevail.

Brown and Wray[143] surveyed the activities of six enzymes (hexokinase, phosphoglucose isomerase, phosphofructokinase, aldolase, glucose 6-phosphate dehydrogenase, and amylase) in extracts of cotyledons from germinated peas. For the first 10 days, activities underwent changes which indicated control of synthesis as well as destruction of those enzymes. With the exception of phosphofructokinase which decreased from onset of germination, the other enzymes reached a maximum activity after 2 to 8 days then declined. Those pea seeds were dry-sown, prolonging imbibition as compared to other studies in which seeds were soaked.

Peas soaked under aerobic, anaerobic, and aerobic conditions in the presence of ethanol or acetaldehyde led to the suggestion that a decrease in alcohol dehydrogenase activity results from germination under increasing aerobic conditions.[144] In cotyledons of germinating chickpeas, phosphorylase plays a significant role during the first 2 days of germination but is relegated to a secondary role as amylase activity increases.[145] The same study reported a steady decrease in simple sugars during the initial 24 h of germination that was followed by an increase during the subsequent 48 h. Finney et al[146] reported similar decreases in simple sugars of mung and garbanzo beans which were germinated for 48 h. Chemical and breadmaking properties of those beans, including the capacity to support yeast sugar metabolism were also studied. Unlike mung beans and other legumes previously tested, 48 h germination reduced yeast gas production in whole garbanzo bean/water slurries, indicating a net loss of fermentable sugar.

In chickpeas, carbohydrate metabolism during germination appears to be separatable into two phases, one involving glycolytic enzymes and another, appearing after 24 h in which glycolytic activity, including alcohol dehydrogenase, decreases and aerobic processes appear.[143,144,147]

The effects of germination on carbohydrate constituents in seeds are influenced by many factors such as the amount of oxygen and other constituents in the steep medium, the temperature, and the procedure of hydration from dry seed. Those factors will profoundly influence respiration, and breakdown and synthesis of seed carbohydrates.

Germination and Oligosaccharides

Nigam and Giri[148] were among the first to study the effects of germination on the non-reducing oligosaccharides, raffinose, stachyose and verbascose of Indian legumes. In seven major edible legumes the sugar composition was consistent, sucrose, 1.3 to 2.7%; raffinose, 0.4 to 1.1%; stachyose, 1.8 to 2.7%; and verbascose, 3.1 to 4.2% of whole seed weight. Upon germination, green gram lost approximately 1/2 of each oligosaccharide in 24 h. After 48 h of germination 75% of verbascose, 85 to 90% of stachyose and 100% of raffinose was utilized. Moreover, the sucrose content decreased by about 20%. Fructose increased from a negligible value to 6.0 g/ 100 g seed. The rapid interconversion of those saccharides is indicative of their importance in germination.

A rapid hydrolysis of raffinose and stachyose also occurred in germinating soybeans.[149] In 2 days, about 50% of the oligosaccharides were utilized. After 4 days about 80% of those sugars had disappeared and by 6 days none could be detected.

Subbulakshmi et al[150] reported similar effects of germination on starch, and reducing and non-reducing sugars in horse gram and moth bean. After 72 h of germination, total soluble sugars increased from 13.8 to 17.5% and 15.0 to 20%, respectively. During that period total reducing sugars increased from 6.7 to 10.7% and from 3.5 to 13.8%, respectively; while non-reducing sugars from the two legumes decreased from 7.1 to 6.8% and from 11.5 to 6.2%, respectively. During that period the starch content of horse gram was reduced from 39.6 to 34.7% while that of moth beans was similarly reduced from 43.7 to 40.5%. Ganesh Kumar and Venkataraman[151] also germinated three other legumes (chickpeas, cowpeas, and green gram) for 72 h and reported similar results: reducing sugars increased from 3.0, 4.5 and 2.5% to

14.7, 13.6, and 17.0%, respectively; while non-reducing sugars decreased from 13.3, 15.5 and 17.5% to 6.6, 3.9, and 6.8%, respectively. During 72 h of germination more starch was depleted than reported for horse gram and moth bean.[150] Starch from chickpeas, cowpeas and green gram decreased from 40.5, 40.5, and 38.3% to 29.3, 27.0, and 22.5%, respectively.

Others have noted similar effects of germination on galacto-oligosaccharides. Dry mung beans have low glucose (0.4%), fructose (0.4%), and sucrose (1.6%) content but a significantly greater quantity of galactose-containing sugars (3.9%).[75] Three days of sprouting increased glucose and fructose 10-fold, sucrose doubled and galactose-containing sugars were depleted. Germination also depleted non-reducing sugars in gram and peas.[152] Germination for 72 h brought the raffinose, stachyose and verbascose content of mung beans and chickpeas to nil.[153] Four days of germination exhausted the raffinose and stachyose content of blackeye and pinto beans.[154] The sucrose content of those legumes rose to a maximum after 5 days of germination and then declined, while fructose continued to increase even after 6 days of germination. Gupta and Wagle[155] reported that 96 h germination essentially depleted the raffinose and stachyose content of black gram and mung bean.

Effect of Germination on Other Carbohydrates

Germination also affects carbohydrate constituents other than those already discussed. Azhar et al[156] showed that starch and fructosan levels in chickpeas fell from 530 and 11.0 mg/g to 275 and 4.6 mg/g after 168 h germination. Red kidney bean starch declined from 59.1 to 37.0% after 4 days germination and to 7.2% at 10 days.[154] Free sugars (as glucose) increased from 1.3 to 3.2% after 4 days, and to 11.4% after 10 days. Per Åman[153] studied the effects of germination on the rhamnose, fucose, arabinose, xylose, mannose, galactose, glucose and uronic acid content of mung beans and chickpeas. There were minor changes in those sugars during germination.

Twelve days germination increased free pectin of green gram and cowpeas from 6.2 to 11.9 meq/100 g and 6.7 to 8.8 meq/100 g, respectively.[124] Free pectin of chickpeas decreased from 3.4 to 1.9 meq/100 g, after 72 h germination. Cooking increased free pectin of green gram and cowpeas but did not alter that fraction in chickpeas whether germinated or ungerminated.

The pattern of starch degradation in storage parenchyma of the pea, like that of protein, begins with degradation at the periphery of cotyledon and progresses inwardly. After two days small starch grains (3-15 μ) appear in the storage cells and aggregate near the nucleus. They disappear from the outer region of the cotyledon by day 5 and from the center core at 7 to 12 days.[170] In bean, cotyledonary nitrogen and starch decreased about 10% and 60%, respectively, after 4 days germination. Soluble sugars increased 275% in the same period. However, by day 7 soluble sugars were approximately 15% lower than in the original seeds.[172]

As measured by starch paste properties, the degradation of starch in 4-day germinated peas, lentils and faba beans was correlated with an increase in α-amylase activity.[312] Lentils and peas gave pasting temperatures greater than did faba beans. Germination effected the former and it also had more deleterious effects on the bread-making properties of lentil and pea starches as compared to bean starch.

Flatulence

Flatulence associated with high legume consumption is regarded by many investigators as due to anaerobic microbial degradation of certain oligosaccharides which mammals cannot otherwise utilize. The three principal oligosacchardies of mature legume seeds are raffinose, verbascose, and stachyose in which galactose is present in α-linkage. Two to four days of germination virtually depleted those reserves. Here, consideration will be given to the presence or absence of oligosaccharides and their relationships to the problem of flatulence.

Calloway et al[157] studied the reduction of intestinal gas-forming properties of legumes that had been prepared by traditional and experimental food processing methods. That involved evaluation of flatus and measurement of bacterially induced gas in the exhaled breath of healthy young male volunteers. Earlier studies showed that "flatulent test meals fed in the morning cause elevation of breath hydrogen, flatus volume, and flatus hydrogen and carbon dioxide content 5 to 7 h later." Mature dry lima beans equalled California small white beans in flatulence-inducing factor. Mung and soybean produced 2/3 the flatus of lima and white beans. Soybean and mung bean sprouts retained most of the factor present in the original bean. Neither peanuts nor tofu (a processed soybean curd) caused flatus. A toasted soybean grit food caused as much flatulence as an equal weight of soybean carbohydrate fed as whole beans. An enzymatic treatment designed to hydrolyze raffinose and stachyose had a neglegible effect on flatulence. Tempeh, a product of mold fermented soybean grits, caused gas production similar to the control and gave a significant delay in gas formation. Ethanol extraction of white beans failed to remove all the flatulant factor but offered promise for future development. Narayana Rao et al[158] studied flatus in 9 to 11 year old females that were fed relatively high levels (30 g/day) of certain legumes. Controls released 31 ml gas/hour and those fed chickpeas, black gram, red gram, or green gram 93, 82, 73, and 62 ml gas/hour, respectively. Both studies found green gram to be the least flatus producing legume among foods tested.[157,158] In rats fed the same four legumes, intestinal flatus ranged from 2.04 to 1.28 ml as compared to 0.59 ml/gas in the casein-induced control group.[159] The carbohydrates of those four legumes were less digestible than corn starch and their order of digestibility was the inverse of their gas inducing capacity (red gram > chickpeas > black gram > green gram).[159,160] Thus green gram produced the least gas and was the most digestible while red gram and chickpeas were the greatest gas inducers and least digestible. All legumes in that study contained similar quantities of stachyose, raffinose and verbascose, yet the flatulence production of green gram was notably lower. Furthermore, 2 day germinated chickpeas and green gram reduced flatulence 10 to 20% while 50% or more of the glacto-oligosaccharide content disappeared in the same period. Thus, other factors contribute to flatus production.

The lack of relationship between oligosaccharides and flatus production is further reinforced by a study of 72 h germinated green gram and cowpeas.[161] Albino rats fed those germinated legumes at a 15% protein level showed no reduction in flatus while an increase was observed when chickpeas diet were supplied. A control diet of casein induced less total gas than any legume but caused more stomach gas than the other diets. The major constituent in casein-induced stomach gas was nitrogen whereas carbon dioxide was the major constituent of flatus in rats on legume diets. Cooking reduced rat flatus from green gram and chickpeas by 35 and 20%, respectively, but slightly increased it in a cowpea diet. In a separate study cooking did not significantly alter flatus that was induced in rats fed green gram or chickpeas.[159] A significant increase in oligosaccharide content as the result of cooking was noted for each of the 4 legumes considered here.[162]

Germination of those four legumes resulted in virtual depletion of raffinose, stachyose and verbascose.[162] The oligosaccharide content of red gram, chickpeas, black gram, and green gram varied by as much as 300, 225, 300, and 145%, respectively, for verbascose plus stachyose in different varieties and by as much as 700, 375, 400, and 350%, respectively, for raffinose.

Jaya et al[163] studied the effect germinating green gram and chickpeas had on the rate of *in vitro* gas production by *Clostridium perfringens*. Germinated legumes produced less flatus than ungerminated ones. Chickpeas produced significantly more flatus than did green gram, a result consistent with other reports. In the same study different fractions from green gram and chickpeas, specifically the starch and ethanol soluble fractions, were tested for flatus production. Chickpea starch induced the most flatus, while the ethanol soluble fraction of green gram was responsible for the most gas produced by *Clostridium perfringens*. The report concludes that in addition to oligosaccharides, other carbohydrate constituents and certain ethanol insoluble fractions contribute to flatulence.

FATS AND LIPIDS

Germination of chickpeas was accompanied by quantitative changes in lipid constituents.[164] The decline in the concentration of mixed tocopherols, phytosterols and phasphatides exceeded changes in total acetone soluble lipids. During steeping, changes in those constituents accounted for nearly 50% of total change, an indication that even during steeping chickpeas activate appropriate enzymes which then initiate a complex series of lipid linked metabolic reactions. During the 25 h steep stage 95% of the phosphatide-bound choline was utilized. Later Talwalker[165] clearly demonstrated the presence of phosphatidylcholine phospholipohydrolase (phospholipase D) activity in germinated chickpeas. No activity was found in ungerminated seeds. It was not clear from Talwalker's study whether the enzyme activity was due to *de* novo synthesis or due to activation of the enzyme or liberation of the enzyme from some inhibitor.

Three days germination of alfalfa, lentils, mung beans and soybeans decreased fat content from 13.6, 1.8, 1.6, and 21.9% to 5.1, 1.1, 1.4, and 9.7%, respectively.[75] In wheat, crude fat was increased by 3 days germination from 0.95 to 1.54%.[166] The fat content of de-rooted mung bean sprouts and whole bean prepared by Chinese methods was about 0.9%.[288] Total lipids in wheat increased from 2.89 to 3.37% after 4 days sprouting.[167] The 16% apparent increase resulted primarily from loss in dry weight due to respiration and leaching. A study on the effects of 10 days germination on the fat content of normal and high lysine sorghums revealed no change in fat content of normal sorghum but a moderate increase in the high lysine variety.[168]

PROTEIN

Protein Differentiation and Synthesis

Young and Varner[141] studied enzyme synthesis in young pea seedlings and concluded that there was no significant change in the level of protease activity after 8 days germination. In contrast, amylase and phosphatase activities increased many fold with amylase increasing to the end of 8 days and phosphatase maximizing after 5 days germination. By the use of protein synthesis inhibitors it was shown that the increase in phosphatase activity was almost certainly the result of new protein synthesis. "The intra cellular distribution of this enzyme indicates that the synthesis occurs in the microsomal fraction. The phosphatase has been purified several fold and shown to be highly specific for ATP and ADP and to be maximally activiated by cadmium ions."[141]

In germinating peas increases in alcohol-soluble and α-amino nitrogen were accompanied by a decrease in the total nitrogen of the cotyledons.[169] Two casein hydrolyzing systems were detected, one with a pH optimum of 5.5 and the other, 7.0. It was suggested that those two enzymes selectively hydrolyze the principal protein reserves, legumin and vicilin.

Smith and Flinn[170] described the pea cotyledon as being "bounded by an epidermis from which stomata are absent except in the region adjacent to the embryonic axis. Below the epidermis is a single row of cells comprising the hypodermis. The storage parenchyma makes up the bulk of the cotyledon. It appears to consist of two distinct zones, an outer zone corresponding to the spongy parenchyma of the leaf and an inner zone corresponding to the palisade tissue. Both zones are much less distinctive than the corresponding tissues in the leaf. The cotyledon possesses a well developed vascular system which runs more or less along the boundry of the inner and outer storage tissues. In the mature embryo it consists entirely of procambium: no native xylem or phleom elements have been detected in it. Differentiation of vascular tissues from the procambium occurs during the first two days of germination." Differentiation of phleom precedes that of the xylem, with differentiation of the phleom first observed 12 h after soaking and the xylem about 4 h later.[170] Within 24 to 32 h a continuous conducting system throughout the cotyledon was present and within 48 h after initial steeping differentiation was complete. At day one, higher levels of RNA were present in the epidermis, hypodermis, and some storage cells than prior to germination and by day 9 stored reserves were completely exhausted.

During the first two days of germination, the protein bodies in the parenchyma appeared to enlarge to at least twice their original diameter and subsequently to fuse together in groups to form larger aggregate bodies, which ultimately disintegrated, and to form small, irregular bodies which finally disappeared. Protein bodies in the embryo of Yucca schidigesa coalesced before breaking down but those in the perisperm appeared to breakdown directly.[171] After about 5 days the protein disappeared from the outer layers and after about 12 days the storage parenchyma was nearly devoid of protein with only the nucleus protein remaining. Protein body degradation in the cotyledon occurred first in the epidermis and in cells adjacent to vascular bundles,[172] as noted also by Smith and Flinn[170] and Tombs.[173] By contrast, in peanuts and common beans, breakdown was slowest in epidermis and vascular bundles, respectively.[174,175]

Marcus and Feeley[176] reported that ribosomes from wheat embryos were capable of protein synthesis 16 h after initiating imbibition. Polysomes were the only ribosomal fraction capable of amino acid incorporation.[177] The time required for ribosomal activation and polysomal formation was apparently within one half hour after imbibition was initiated.[178]

Germinating chickpea protein was depleted in cotyledons and accumulated in developing seedling along with increases in amino acids.[156] The net proleolytic activity of the growing tissue was 7-fold that of the cotyledons. That activity in the seedlings presumably arose by synthesis of new proteins. Earlier Pal and Pal[289] reported sprouting of chickpeas increased nitrogen and protein and that atmospheric nitrogen was alone responsible for the increase. Hurst and Sudia[290] reported soybeans germinated in light lost about the same dry weight by day 8 but took up more nitrogen from nutrient solution than those sprouted in the dark.

Yomo and Taylor,[179] who studied common beans during early stages of germination, observed protease activity throughout the cotyledon except in two or three cell layers below the cotyledon surface and in several cell layers around the vascular bundles. The source of the protease appeared to be a highly active cell layer surrounding the protease-inactive cells near the vascular bundles.

Mandal et al[121] isolated and purified a phytase from mung beans germinated 72 h. It had a 7.5 pH optimum at 57°C (M.W. ≅ 160,000). A phosphatase present in phytase preparation had optimal activity at pH 6.0. The pH optimum of wheat phytase was 5.1.[107] Eighty to ninety percent of the protein in the cotyledonary fraction of ungerminated chickpeas was salt soluble globulin comprising two distinct fractions.[180] Cerletti et al[181] were able to extract about 13% of lupine protein with water at pH 5.0. An additional 77.4% of the protein was extracted with 1 N NaCl. At pH 7.0 after water extraction, 0.1 N NaOH extracted 93.3% of the protein but the solution was unstable and after 18 h (4°C) only 38.7% protein remained in solution.

Swain and Dekker[182] found an α-amylase in cotyledons of germinating peas with a specific activity comparable to the most highly purified crystalline preparation of that enzyme. Compared to a barley malt which crystallized after being purified only 65-fold, a purification of at least 3400-fold was required for the amylase from pea cotyledons. The optimum pH for β-amylase from pea stems was 5.2 to 6.1, typical for β-amylases in plants.[183] Kumar and Venkataraman[184] noted increases in amylases of about 50% for chickpeas and cowpeas and 100% for green gram after 72 h germination.

Amino Acids

Heller[40] analyzed ungerminated mung beans for cystine, histidine, argenine and lysine using chemical methods. Rats were used to show that mung bean protein was nutritionally adequate to maintain rat growth to maturity but with limited reproduction.[291] Belton and Hoover[185] compared microbiological and chemical methods for assay of amino acids in mung beans and reported reasonable agreement. They recalculated Heller's values[40] in terms of amino acids as percent total protein, and reported that his results compared favorably with theirs. Bagchi et al[186] examined the amino acid contents of 8 Indian pulses and found all to be devoid of β-alanine. Either histidine or agrinine, but not both, was also lacking. In addition only chickpeas lacked cystine, methionine and tryptophan. In contrast to that, Vijayaraghavan and Srinivasan[187] reported that chickpeas contained more methionine than any of five other legumes when evaluated using a colormetric method. Desikachar and De[188] reported that pulses generally contained about twice as much total sulfur amino acids as cereals. Also, soybeans and sesame seeds contain about 4-fold more cystine and twice as much methionine as cereals or about 3-fold more total sulfur amino acids as most cereals.

Everson[189] found no increases in essential amino acids in soybeans but did find that germination increased nutritive value. Wu and Fenton[33] noted a loss in total lysine content in 6 to 7 day sprouted soybeans. In 1962 Cook[190] made reference to lysine biosynthesis in germinating barley. Tsai et al[191] noted an increase in lysine of 50% in maize embryo. A similar increase in tryptophan in the embryo after 5 days germination was also found. Germination increased lysine in wheat, triticale, barley, oats and rye.[192] Wheat increased the most (about 50%), other cereals about 10 to 35%, with the exception of rice which failed to gain in lysine. Tryptophan increased about 20 to 25% in wheat and rye while in the other cereals it remained the same when dry weight losses were taken into consideration.

Wheat varieties differ in rate of lysine biosynthesis during germination.[193] During 7 days germination, lysine increased 30% in a Fortuna composite and 50% in a single seed lot of Nugaines. Forty-two h germination of black gram gave maximal production of lysine.[35] Lysine increased from 4.10 to 5.21 g/16 g N, a 27% increase. Dry weight losses were no more than 10% during the germination period. Vitamin C also reached a maximum at the same time (≅ 18.1 mg/100 g), then rapidly decreased.

Hsu et al[37] developed an automatic rinsing, controlled temperature and humidity cabinet with vigorous air movement which produced a high degree of sprouting in yellow peas, lentils, and faba beans (e.g. average epicotyl + hypocotyl length of yellow pea was more than 9 cm after 4 days germination). Nevertheless authors noted very minor changes in the 18 amino acids after the 4 days vigorous sprouting. Finney et al[194] used the same equipment to study faba bean germination and found relatively minor changes in amino acid distribution. Aspartic acid, ammonia and serine increased 17.2, 10.8, and 7.2%, respectively, whereas isoleucine, tyrosine, glycine, and arginine decreased 11.8, 9.5, 7.2 and 7.0%, respectively. Pomeranz et al[292] found only minor differences in the amino acids of germinated and ungerminated soybean flours. Germination of normal and high lysine sorghums increased lysine content of normal sorghum from 2.2 to 3.2 g/16 g N after 10 days and high lysine sorghum from 3.0 to 7.8 g/16 g N.[168] There was a large increase in albumin (rich in lysine) and a large decrease in kafirin and cross-linked kafirin (both low in lysine) during sprouting.

Twelve wheat variety seed composites were germinated and analyzed for complete amino acids.[129] After 7 days germination dry weight losses varied from 24 to 40% while total weight gain varied from 390 to 575% among the 12 composites. After an initial rise in lysine content that reached a maximum in 4 to 5 days, lysine decreased for all 12 varieties. Absolute increases in lysine at 5 days varied from 5% to more than 80% for different wheat varieties. In addition to the increase in lysine there were significant changes in other amino acids, and for 10 varieties both dry weight loss and total weight gain was related directly to most changes in those amino acids. Glutamic acid which accounted for 28 to 31.5% of the amino acids in the ungerminated 12 varieties decreased to a value of 11 to 21% after germination. Proline was 9 to 11% before and 4 to 8% after germination. The corresponding values for aspartic acid were 4.8 to 6.0% and 15 to 25.5%.

Supplementation with Amino Acid

By 1920 Johns and Finks[195] had established that cystine was essential for normal growth when navy beans were the principal source of protein in foods. Basu et al[196] found that lentils contained much less cystine than green gram and correlated the lower biological value of lentils to its lower cystine content. Rat feeding studies confirmed the superiority of green gram to lentils and showed that it was nearly as effective in promoting normal rat growth as a diet of milk and whole wheat.[197]

A number of studies have demonstrated that the protein efficiency ratio (PER) of nearly all legumes is improved when supplemented with as little as 0.1 g/100 g methionine.[198-202] Higher levels of methionine brought no added improvement. Soybeans and chickpeas were far superior to other legumes when fed to rats without amino acid supplementation.[198] The PER of the soybeans and chickpeas increased from 2.1 and 1.8 to 2.8 and 2.7, respectively, when the basal diet was supplemented with 0.1% methionine.

When Phansalkar et al[203,204] replaced 10% of the protein in a rat diet with leafy vegetable, only amaranth increased the PER. Appreciable improvement in PER was observed in diets when 30% of the protein was from pulse and 70% from cereal. Rice was an exception. When the vitamin and salt mixtures were omitted from the cereal:pulse:amaranth diet, no change in PER was observed. A review of the major legumes, cereals, and nuts singled out chickpeas as the superior legume with the highest PER.[205]

In most legumes, methionine is the limiting essential amino acid, followed by tryptophan.[132] The effect of those limitations is seen more markedly on growth and less on protein requirements for maintenance. The lysine content of legume protein varies from about 80 to 150% of whole egg protein and is one reason why legumes are an excellent choice when fed in conjunction with cereals all of which are deficient in lysine.[132] Autoclaved navy beans supplemented with about 0.2% methionine promoted as much growth as a casein control diet.[206]

Effect of Germination on Protein Quality

Soybeans germinated 60 h compared favorably with immature soybeans and were distinctly superior to mature soybeans when fed either raw or heated in a 10% protein diet to rats.[207] PER values for raw-mature, -immature, and germinated feeds were 0.48, 1.10, and 1.30, respectively, and for cooked-mature, -immature and -germinated feeds, 1.71, 1.90 and 2.01, respectively. Chickens fed germinated soybeans improved mean live weight from 167 to 194 g/5 weeks but only marginally improved PER from 0.28 to 0.30.[208] That study also tested rats and confirmed results of Everson et al[207] in that sprouted soybeans nearly doubled weight gain. Sprouting materially improved PER as well.

Chattopadhyay and Banerjee[209] germinated 5 popular Indian pulses and compared their feeding values to a casein control in young albino rats. Germination significantly increased feeding value of chickpeas but did not significantly change the PER values of black gram, lentils, mung beans or peas.

Fenugreek leaves and seeds are widely used as both spice and herb, as well as a main food in many parts of the world, including Egypt, Southern Europe, and India.[210,211] The seeds are reported to possess pharmacological properties and contain relatively large amounts of protein and Fe, more than average amounts of P, Ca, and carotene, and have been reported to accelerate growth in rats. Raw and germinated seeds are rich in lysine and germination is known to increase methionine. Rajalakshmi and Subbulakshmi[212] compared a rice:black gram diet (50:50) for rats, with diets in which germinated or ungerminated fenugreek substituted for 10% of the black gram against a control diet of rice:skim milk (72:28). Ungerminated fenugreek improved the rice:black gram diet and compared favorably with rice:skim milk control diet. Germinated fenugreek diet significantly improved both mean final weight and protein efficiency.

Jaya et al[213] fed rats raw or cooked, ungerminated or germinated green gram, cowpeas or chickpeas 10% protein diets. Rats fed chickpea diets gained more than twice as much weight as rats fed either green gram or cowpeas, regardless of the legume processing. Germination improved the chickpea PER but not those of green gram or cowpeas. Heating the meals improved all preparations with respect to weight gain and PER.

Twenty-four h germination appeared optimum for chickpea effect on growth and PER.

No significant differences were found in PER of rats fed green mature, dry mature, or 4-day germinated soybeans.[214] Cooking increased all PER values from about 0.68 to 2.65. Khan and Ghafoor[215] found that soaking 12 h or soaking and germinating 48 h of black gram increased and then decreased PER (0.4, 1.9 and 1.1, respectively) and weight gain (11.5, 68.5 and 33.0 g, respectively) of rats fed 10% protein diets.

Germination of corn and sorghum increased relative nutritive values and the content of lysine, methionine and tryptophan.[293] While a 50:50 blend of germinated and ungerminated corn was acceptable in tortillas, it was less desirable than tortillas made from ungerminated corn.[294] Both germination and fermentation (separately) improved the relative nutritive value of a number of cereals.[295] Germination and fermentation also significantly increased available lysine in a number of cereals. Hasim and Fields[296] reported germination increased niacin and riboflavin content of corn and the relative nutritive value as measured by *Tetrahymena pyriformis* W. Processing germinated corn meal into corn chips decreased the vitamins and relative nutritive value but they remained higher than the ungerminated control.

Jaya and Venkataraman[216] evaluated the effect of germinated and ungerminated chickpeas or green gram in rice or wheat diets at the ratio of legume to cereal protein of 1:2. Both germinated and ungerminated chickpea and green gram flours significantly increased rat weight gain when supplementing rice or wheat. Chickpea flours always increased PER more than did green gram. Two days germination significantly improved the supplementary value of chickpeas when combined with rice, however germinated and ungerminated chickpeas equally improved PER when combined with wheat. Green gram flours (germinated or ungerminated) failed to improve PER when combined with rice but both did equally well when supplementing wheat. PER values of chickpea and chickpea:cereal flours were invariably higher than were green gram and green gram:cereal flours.

Digestibility and Biological Value

Bhagvat and Sreenivasaya[217] reported that when preparing globulins and albumins from pulses "there exists in the saline extract a nitrogenous fraction, nonprotein in character, diffusible through dialysing membranes, non-precipitable by saturation with salts, non-coagulable by heat and indifferent to drastic protein precipitants like trichloracetic acid." That fraction contains a high proportion of peptides which is very much like the portion often given in India to invalids and children, especially during convalescence. It was noted that the average complexity of the fractions from various pulses was related to the recognized ease of their digestibilities. Phaseolus aconilifolius and Vigna cylindrica (blackeyed peas) were indicated as having the most "non-protein" nitrogen, while chickpeas had the highest ratio of that fraction to total protein, indicating the former were most digestible while the latter was the least digestible, a fact which is in harmony with common Indian understanding of those pulses when prepared in the ungerminated state. Swaminathan[218] used a copper hydroxide method to find that non-protein nitrogen varied from 2.91 to 14.5% of total nitrogen for various Indian legumes while it varied from 1.9 to 7.0% for cereals. The average values for legumes and cereals was 9 and 5%, respectively, while milk averaged 9% non-protein nitrogen.

Although chickpeas (bengal gram) are known to contain more indigestible components than many other common legumes, their nutritive value is recognized as comparatively high, as demonstrated by Vijayaraghvan and Srinivasan[187] who analyzed the essential amino acid contents of 6 popular Indian legumes and calculated the relationship between the chemical scores, essential amino acid contents and biological values (Table 10). It will be noted that chickpeas had the highest values. Of the ten most popular Indian legumes, chickpeas were most nutritious, having the highest biological value and net protein value. It ranked fourth in digestibility. Mung bean ranked second among the ten legumes.[297-300] A review of the assigned biological values of the most popular 15 Indian legumes varied from 32 to 78, with various sources giving a rather wide range for each legume.[132,205] Generally chickpeas were reported as having the highest biological value but were somewhat less digestible than peas, lentils, and green gram.[205] The digestibility coefficient of legumes varied a great deal, with values between 51 and 92% being reported.[132] Cooking and autoclaving did not greatly enhance digestibility coefficients of most legume proteins.

Table 10. Correlation of chemical score, essential amino acid (EAA) index, biological value and digestibility coefficient of pulses.[187]

Food Protein	Chemical Score	EAA Index	EAA Index Less Arginine	Biological Value (BV)	BV X DC / 100	Calculated Biological Value
Bengal gram	30.3	67	63	78	59	58
Red gram	13.4	67	66	74	46	45
Green gram	25.9	72	70	64	53	53
Black gram	28.3	71	70	60	50	55
Cowpea	26.7	72	69	57	33	54
Lentil	23.5	60	56	58	45	51

Since Ca is notably low in wheat, rice and most cereals, ragi or millet which has a greater content of Ca and methionine than any other common cereal, is particularly important since it is consumed in large quantities by the poorer classes in many arid parts of the world, including Sudan and India.[219] Ragi when mixed with either black-, red-, green-, or bengal-gram (chickpea) plus 5.2% dry milk solids maintained young rats nearly as well as ragi and high levels of dry skim milk protein, with their biological values ranging from 84 to 89% while ragi plus a high level of skimmed milk was 94%. Addition of skimmed milk to ragi plus one of six pulses improved the biological value from an average of 76.6 to 88.2%.[132]

Seventy-two h germination did not improve the biological value of green gram, cowpeas, or chickpeas, and only green gram increased in digestibility (from 72.4 to 81.2%) after germination.[220] When legumes were subjected to *in vitro* amylolysis it was found that 72 h germination increased carbohydrate digestibility. Ten days germination of red kidney beans increased protein digestibility of both raw- (29.5 to 64.4%) and cooked-bean (69.3 to 84.4%).[221]

Black gram soaked 12 h increased protein digestibility from 72 to 90% but germination (48 h) reduced it to an intermediate value of 81%, without essentially changing biological values.[215]

Subbulakshmi et al[150] evaluated the effects of germination on dialysable nitrogen, non-protein nitrogen and *in vitro* digestibility of horse gram and moth bean. There was a progressive increase in non-protein nitrogen accompanied by a decrease in protein nitrogen. After 72 h germination 25% of the nitrogen was present as nonprotein nitrogen. After 72 h germination moth bean protein digestibility increased from 69 to 94%, while that of horse gram increased from 80 to 87%. The rates of starch digestiblity (α-amylase) for moth bean and horse gram were similar with 72 h germination increasing both between 20 to 35% and cooking increasing each about 2.5- to 3-fold.

Subra Rao and Desikachar[222] reported that the indigestible residues in feces of rats and humans were greater if fed chickpeas, black gram or soybeans than if fed green gram. Also feces of rats fed green gram were smaller and less bulky than when fed chickpeas or soybeans. However, growth rates of rats fed on chickpea and soybean diets were reported to be greater than if fed other pulses, including green gram, showing that the actual nutritional value was not effected by the larger indigestible residue which remained in fecal bulk. Similarly, authors found that indigestible residues of pulses fed 16 to 17 year old male humans were similar to the rat studies.

Protein Inhibitors

The common cereals contain little or no trypsin inhibitor activity or other anti-nutritive factors. Although many legumes do contain such factors, many do not. Borchers and Ackerson[223] investigated trypsin inhibitors of 30 legume and 9 cereal species and found 13 of the legumes and all of the cereals lacked this activity. Among the remaining legumes 6 contained very little trypsin inhibitor (including golden mung beans and alfalfa) and 11 contained substantial quantities. The latter included soybeans, blackeye peas, lima beans, chickpeas, and carab beans. Later Borchers and Ackerson[224] fed seeds of 11 legume species, raw or autoclaved, to rats as the sole source of protein at 12% level. The nutritive value of 5 species was not improved by autoclaving.

Trypsin inhibitor activity in different species is variably inactivated depending on the type of heating as well as the temperature. For example, Osborne and Mendal[225] reported dry heat was appreciably less effective than wet cooking heat in improving the feeding value of soybeans. Since then other workers have shown that initial soaking prior to autoclaving was required to completely eliminate toxic factors in kidney beans,[226] soybeans,[227] and field beans,[228] but was not required in finely ground navy beans.[206]

The effects of germination on nutritive value and anti-nutrient factors in edible legumes are also variable and reports are sometimes in conflict. Germination improved the nutritive value of soybeans.[188,189,208,229] Disikachar and De[188] reported that germination left proteolytic inhibitor activity unaltered. More recently Bates et al[214] reported that 4 days germination reduced trypsin inhibitor activity of soybeans to 33% of the amount present in green-mature or dry-mature beans.

Two days germination did not alter the trypsin inhibitor content of chickpeas, black gram, lentils, mung beans, or peas and only improved the nutritive value of chickpeas.[209] In contrast, Devadatta et al[230] reported that germination did not improve the nutritive value of chickpeas. Soaking navy beans 1 to 4 days reduced trypsin inhibitor and hemagglutin activities but prolonged germination effected only trypsin inhibitor activity. Germination had no effect on nutritive value of the beans.[231]

Palmer et al[232] ruled out trypsin inhibitors as the main toxic component of raw kidney beans. Subbulakshmi et al[150] reported that 72 h germination decreased hemagglutinin activity of horse gram from 2.6 to 0.6 H.U./g seed X 10^3. Moth bean possessed essentially no hemagglutinin activity. Trypsin inhibitor activity of horse gram decreased about 20% while that of moth bean decreased nearly 60% after 72 h germination. Heating inactivated trypsin inhibitor activity in both legumes.

Ten day germination of red kidney beans (sprouts only 3-3.5 cm in length) lowered trypsin inhibitor activity about 50%. Since the globulin fraction contained more than 10-fold the trypsin inhibitor activity of whole beans, yet was more digestible than whole beans, additional factors were responsible for the low digestibility of raw beans.[233] Bau and

Debry[36] found germination reduced trypsin inhibitors by 30% and improved nutritional quality (PER) of some soybean protein products.

Gupta and Wagle[234] reported that during germination of several Phaseolus species the trypsin inhibitory activity decreased by 40% after 9 h then began to increase to 300 to 350% of the original activity after 72 h germination. Beyond 3 days, trypsin inhibitory activity decreased and on day 9 germinated seeds contained 50% of their original amount. A similar pattern was found for lysine biosynthesis in 12 wheat varieties over 7 days sprouting by Finney et al.[129] Both studies illustrate the importance of length and degree of sprouting and dramatically show how many biochemical changes associated with germination can fluctuate even during the first few days or hours.

The physiological function of the so-called antinutrient factors and their role during early germination of edible seeds is uncertain. Pusztai[235] demonstrated that kidney bean contain several proteins which inhibit a number of mammalian proteolytic enzymes such as trypsin, chymotrypsin, elastase, and plasmin. A number of those protein inhibitors have been purified and are known to be low molecular weight proteins with as much as 15 to 20% cystine content.[236-238] With such a high content of cystine, and because of the substantial amounts of those trypsin-inhibitory proteins in many edible seeds, Pusztai[235] reasoned that their high cystine content might be related to a role as a cystine and/or sulphur depot protein. The high cystine content is notable when considered in the light of the known deficiency for sulfur-containing amino acids of bean protein. Freed and Ryan[239] pointed out that trypsin inhibitors are not rapidly decreased during germination therefore it is unlikely that they function as storage protein in soybean seeds.

Pusztai[236] isolated a trypsin inhibitory protein with about 15% cystine from red kidney bean. It was partially degraded during germination. In the case of some legumes cystine may be released during germination thereby sparing the low methionine content. Kwong and Barnes[240] provided evidence that adequate cystine spares the need for methionine in soybeans by as much as 40%. They found that when the rat is fed unheated soybeans, methionine utilization is not impaired. Rather, there is a metabolic block in the utilization of cystine for protein utilization, and that block is somehow caused by the trypsin inhibitors of the unheated soybean.

Hobday et al[241] investigated the subcellular distribution of protease activity, trypsin-inhibitor activity, and the effect of the inhibitor on the protease, and found no evidence that trypsin inhibitors regulated proteases during germination of peas. Thus the precise role the protease inhibitors play in seed physiology remains to be clarified. Additionally no clear relationship has been established linking inhibitor activity, ease of inhibitor inactivation, and feeding value of the various legumes.

Polyacrylamide gel electrophoresis patterns of legume protein from yellow peas and lentils were altered during 4 days germination but those from faba beans showed little change.[313] The emulsifying capacity and solubility of those protein fractions were unaffected by germination. Foam forming capacity increased and foam stability decreased in the germinated samples. Ungerminated or germinated faba beans were superior to peas or lentils in bread-baking properties.

HUMAN FEEDING STUDIES

Evidence to indicate that sprouting of edible seeds promotes nutritive value as measured by chemical composition, biological value and animal feeding trials is found throughout this chapter. The ultimate test for evaluation of nutritive benefits derived from sprouting cereals and legumes for food use rests, however, on controlled human feeding studies. There are some well docmumented cases, all of which substantiate the chemical evaluations and animal feeding studies.

Treatment of Scurvy and Malnutrition

Wiltshire[7] compared 27 severe cases of scurvy fed 4 oz per day of 3 day germinated Haricot bean with 30 cases fed 4 oz fresh lemon juice. Beans outperformed juice with the result that 70% versus 53% of the cases were cured within 4 weeks.

Scurvy and malnutrition were stamped out by public distribution of germinated grain to over 200,000 people in Northern India during the last phase of a 3 year famine (1938 to 1941).[19] In certain parts of India germinated pulses are a common component of the diet. In South India, sprouted chickpeas are most popular.[31] Sprouted chickpeas were used in child-welfare centers as an inexpensive source of vitamin C.[31]

The biological value of pulses on human subjects was greatest for chickpeas (63.7). Red gram gave a value of 61.7 and mung beans, 43.0.[132] Digestibilities for all three were greater than 90%. The biological values on human subjects for rice or wheat were 50 and 59, respectively.[242] Inclusion of pulse in those diets increased the biological values to 67 and 63, respectively. Basu et al[242] emphasized that one of the greatest defects of the Indian diet was an insufficiency of Ca.

Treatment of Kwashiorkor

Successful treatment of nutritional edema (Kwashiorkor) in children was first recorded in an experiment described in Table 11.[132,243,244] Although the vegetable protein diets were not strictly comparable (there were differences in formulation in addition to the use of germinated versus ungerminated bengal gram), there was a definite tendency for the germinated diet to out perform the ungerminated one.[243] That is noteworthy since animal-feeding studies showed that cereal-legume mixtures out performed either cereal or legume alone. The average number of days required to eliminate diarrhea, often a particularly important aspect of rehabilitation of Kwashiorkor, was less for both vegetable diets than for the casein control diet.

Child Feeding Trials with Germinated Wheat:Bengal Gram

Rajalakshmi and Ramakrisnan[245] studied the effects of feeding scientifically based, nutritionally sound meals and snack foods to young children (average 3-1/2 years old) who were under-fed. Those children suffered gross growth retardation and weighed about as much as average, healthy 1 year old children. In the feeding trials, one main protein calorie dish was _conjee_, a cereal:legume mixture in which millet and chickpea seeds were sprouted, partially air dried, roasted and ground. The cereal:legume flours were then cooked in water for a few minutes to form a thick gruel to which might be added milk and salt. For another meal _dhokla_ was given. That was a fermented cereal:legume mixture with added chopped

Table 11. Effects of feeding skim milk, chickpea:rice (19:25) or sprouted chickpea:banana diets in treatment of kwashiorkor.[243]

	Skim milk	Chickpea (Germ.)	Germinated Chickpea (rice)
No. of cases	49	56	19
Days required for disappearance of Edema:			
Mean	12	13	17
Range	3-33	4-45	4-68
Days to reach minimum weight:			
Mean	7.3	9.5	14.0
Range	0-24	0-51	0-64
Days required to increase body weight 1 lb.:			
Mean	5.4	5.8	5.2
Range	2.3-18.3	1.0-18.0	3.0-7.5
Days required for diarrhoea to disappear:			
Mean	10.5	5.7	8.0
Range	2-23	2-13	3-19
Rise in plasma Albumin 10th day:			
Mean	0.75	0.40	0.20
30th day	1.24	1.04	0.63

greens that was steamed in greased pie plates for about 20 minutes, cooled, cut into pieces and seasoned.

Without going into specific details, those foods were well accepted and tolerated by children. The weight gain and biochemical status of those poorly nourished children after receiving diets based on locally available foods but subjected to simple processing using ordinary culinary procedures were comparable to those of upper class children (Table 12)!

Impact of Germinated Wheat:Bengal Gram 'Take-Home' Supplement on Young Children

The preceding child feeding trials were based on a breakfast and lunch provided at a play center.[245] Often such

Table 12. Changes in the clinical status of 'fed' and control children during the experimental period (percentage incidence).[245]

	'Fed' (n = 25)		Controls (n = 20)	
	initial	final[1]	initial	final
Xerosis of conjunctiva	64	24	50	45
Pigmentation of conjunctive	40	4	40	30
Pale tongue	48	8	50	50
Fissured tongue	20	0	20	15
Adipose tissue-deficient	24	0	25	25
Moon face	48	0	10	10
Diarrhoea	48	8	50	50

[1] Most in whom the symptoms persisted had poor attendance.

control was neither possible nor desirable. Even where it was possible, many children below 2 years of age seldom attended. To meet that problem a take-home supplement of conjee [germinated wheat:bengal gram (2 to 1) plus greens and spices] was supplied to replace breakfast at home and to provide a tea time snack. Two packages of the supplement per day were given to the mothers along with groundnut oil reinforced with vitamin A so that a teaspoon of the oil provided 250 μg of vitamin A. Once again, without going into specific detail, it was found that the children were generally healthier and happier than the controls, and grew more and weighed more after the 8 month feeding period (Table 13).

Rajalakshmi[246] provided some explanation as to why germinated cereal:legume mixtures promoted growth in the child feeding studies. Sprouted millet doubled the riboflavin content and increased the nicotinic acid by 20%. Little loss of those vitamins occurred during roasting (Table 14). Two days sprouting of bajra (millet) or sprouting and roasting increased rat weight gain about 36 or 29%, respectively, over a raw bajra control diet.[136] A wheat:bengal gram diet (4:1) more than doubled the rat weight gains over raw bajra, from 14 to 29 g/2 weeks. Sprouted wheat:roasted bengal gram diet resulted in an additional 3 to 4% increase in rat growth rates (Table 15).

Two very popular and apparently nutritious foods, idli and khaman, were also studied.[247] Idli is made from 2 parts of rice to one part black gram by grinding after initial soaking period of 5 to 10 h, and salting to yield a thick batter. The batter is allowed to ferment over night. Khaman, like idli, is soaked 5 to 10 h, ground wet, and fermented. Both khaman and idli batters are ladled into metal containers and

Table 13. Changes in height and weight (means ± SE) of control and 'fed' children during experimental period.[245]

	Controls (n = 16)		'Fed' (n = 18)	
	initial	final	initial	final
Weight, kg	7.6 ± 0.4	9.4 ± 0.3	7.5 ± 0.3	10.1 ± 0.3
Height, cm	73.8 ± 1.5	79.6 ± 1.2	73.6 ± 1.3	80.3 ± 1.4
Age, years	1.5	2.3	1.5	2.3
Bone age, years	1.4 ± 0.1	2.0 ± 0.3	1.3 ± 0.1	2.2 ± 3.2
$\frac{\text{Bone age X 100}}{\text{chronological age}}$	86.0 ± 4.7	90.0 ± 6.5	96.0 ± 6.4	102.0 ± 3.2
	Number	%	Number	%
Children showing:				
Improvement	6	66	11	85
No change	2	22	2	15
Deterioration	1	12	0	0

Table 14. Changes in riboflavin and nicotinic acid contents of Bajra, with sprouting and roasting.[247]

	Riboflavin	Nicotinic Acid
	mg/100 g	
Raw bajra	0.30	4.0
Sprouted bajra	0.62	4.8
Sprouted and roasted bajra	0.51	4.7

Table 15. Effect of sprouting and roasting on 2 week weight gain of rats.

Experiment	Diet	Weight Gain (g)
I	Bajra, raw	14
	Bajra, sprouted	19
	Bajra, sprouted and roasted	18
II	Wheat + Bengal gram dal(raw)(4:1)	29
	Sprouted wheat + roasted bengal gram dal	30

steamed. The thiamin, riboflavin, nicotinic acid and phytate P contents of unfermented and fermented idli and khaman are listed in Table 16.[247] In the case of idli, fermentation more than doubled thiamin and riboflavin and almost doubled nicotinic acid, while decreasing phytate P by 35%. And in the case of khaman, fermentation increased thiamin by about 40% and riboflavin by 188%, while decreasing phytic acid P by 50%. The effects of fermentation on the idli and khaman bioavailability of the thiamin and riboflavin as found in livers after feeding rats for 4 weeks are given in Table 17. Rats gained 22% more weight on fermented diets compared to unfermented ones. Fermentation significantly increased bioavailability of thiamin, 32 and 19%, and riboflavin, 25 and 37%, in idli and khaman diets, respectively.

OPTIMUM SPROUTING CONDITIONS

Optimum sprouting includes the following conditions: (1) transformation of the most seed constituents into required nutrients (vitamins and essential amino- and fatty-acids), (2) minimal microbiological infestation, (3) production of the most growth in the least period of time while developing sprouts with desirable food properties, and (4) minimal seed nutrient losses and maximal absorption of water soluble constituents from the steep and germination medium.

It is difficult to determine which soaking and germination conditions will produce the most wholesome, most nutritious, finest tasting sprouts with desirable food properties. That task is particularly difficult because each seed species

Table 16. Changes in chemical composition of idli and khaman with fermentation.[247]

	Idli		Khaman	
	Unfer.	Fer.	Unfer.	Fer.
	mg/100 g			
Thiamin	0.21	0.58	0.55	0.79
Riboflavin	0.25	0.54	0.26	0.75
Nicotinic Acid	1.18	2.31	-	-
Phytate-P	174	113	270	135

Table 17. Nutritive value of unfermented and fermented idli and khaman fed to rats.[247]

	Idli		Khaman	
	Unfer.	Fer.	Unfer.	Fer.
Weight gain in 4 weeks(g)	30.0	37.0	37.0	45.0
Thiamin(liver) mg/100g	0.68	0.90	0.94	1.12
Riboflavin	2.75	3.42	2.68	3.66

requires a slightly different set of conditions to promote optimum growth. In addition, for some species the "optimum growth" conditions might not produce the other desired sprout properties. Haffenrichter[301] measured the rates of respiration (CO_2 liberation) of two soybean varieties. Both fluctuated noticeably and without periodicity except that the greatest variation in rate occurred during early development and again just before the plants succumbed to starvation. The author found no direct relation between growth and respiration. Thus "most growth" was not necessarily to be equated with "optimum" sprouting techniques. There are other examples. If high amylase activity is sought, lower than maximum wheat or barley growth temperatures are found to promote higher amylase content during malting.[248-250] Also, the seed moisture content during germination profoundly effects growth and final sprout composition.[249,251-254] Temperature changes during germination also greatly effect nutrient composition of sprouts and growth.[253,255,256]

Furthermore the task of determining the optimum conditions for food-sprout production is today nearly as difficult as it might have been a century ago since in only a few of the hundreds of related references have germination conditions been studied. Some exceptions to that rule are found in the malting literature and in plant physiological and agronomic studies.

Some Temperature Effects on Germination

About a century ago Detmer (cited by Schimper[257]) stated that the "cardinal points" for wheat germination are minimum, optimum and maximum temperatures of 3°, 28.7° and 42.5°C, respectively. Duggar[258] gives the "cardinal points" for wheat germination as 5°, 29° and 42.5°C. Jones[259] found 24°C was most promising for germination and early growth of peas, soybeans, alfalfa and red clover. Percival[260] cited 22°C as the optimum wheat germination temperature while Atterberg[261] gave 17°C as optimum for wheat germination. Others[262,263] believed a lower temperature, 15°C, was most satisfactory. Coffman[264] found all small grains studied would germinate at "ice melting" temperature and suggested laboratory germination temperatures for testing should be lower.

Dickson[265] made an important observation from the food sprouting viewpoint. He noted that in wheat germination temperatures of 8° and 12°C promoted hypocotyl (root) growth before shoots showed much growth whereas at 28°C or higher vigorous epicotyl (shoot) growth occurred with very little root growth. Fewer roots and more vigorous shoots are universally sought in food sprouts.

Wilson and Hottes[263] found 10°, 15°, and 20°C were equally effective in germinating wheat. A sand medium, hydrated to 50 percent saturated level, was most suitable for wheat germination at all temperatures studied. Wilson[266] in more extensive studies, found 10°, 15°, and 20°C very suitable for wheat germination, with 15°C the best. He noted that a substantial decrease in germination and seed vigor occurred between 25°C and 30°C and that higher temperatures were increasingly detrimental. Soybeans as well as wheat were shown to germinate at 10° to 15°C and at 25° to 30°C. At lower temperatures soybeans took more time to yield equal growth. Wilson[266] also reported that temperatures between 10° and 30°C were suitable for wheat and soybean germination. Fungal attack was apparently not a problem in those studies. Edwards[267] studied soybean germination temperatures between 24.5° and 40°C and reported that 33°C was optimum for the maximum germination in the shortest time. Two years earlier Edwards[302] reviewed the temperature relations of seed germination and emphasized Atterberg's studies[303] which showed that germination percentage eventually approached 100% for temperatures significantly lower than the "critical maximum" but the percentage germination was progressively lower with increasing temperatures about the "critical maximum." Earliest seedling growth of _Phaseolus multiflorus_, _Helianthus annua_, maize and peas was 31°C.[268] Certainly temperatures just below the "critical maximum" will vary widely among seed-species, -varieties, and -lots.

Deleterious Effects of Distilled Water Soaking

True[269] was among the first to report the harmful effects of sprouting seeds in distilled water. Kidd and West[270] reported that the effects of presoaking in distilled water prior to planting a number of seed species ranged from highly deleterious in common beans to improving germinability in faba beans. Apparently the improvement or detriment of the soaking was related to the seed coat thickness and its ability to contain seed constituents and limit water imbibition rate. Kidd and West[271] published evidence that pea seed germination was hindered by soaking in water at 20°C for 24 h and still more hindered at 10° and 15°C. The damage was caused by leaching of important "food reserves" from the cotyledons, or by oxygen deficiency and excess accumulation of carbon dioxide, or by both.

Gericke[272] established the desirability to employ a weak solution of inorganic salts as a germination medium rather than distilled water, and showed that numerous solutions would secure excellent results. Tang[273] emphasized that leaching appeared to occur more rapidly when seeds were in distilled water than when they were in a well balanced nutrient solution. Other workers noted similar deleterious effects of soaking common beans in distilled water.[274-277]

A high negative correlation occurred between seed vigor and the exudation of seed electrolytes (leaching) when soaked in distilled water.[278] Abdul-Buki and Anderson[304] found sugar leaching increased with increasing seed age but not with accelerated aging except where viability was severely reduced. The balance of leaching and reabsorption of sugars during soaking differed with age of seeds because the more viable seed utilized and continued to reabsorb the sugar. Perry and Harrison[279] showed that germination of peas was often depressed by soaking, with low vigor seeds depressed more than high vigor seeds. Lower soaking temperatures were an additional detriment to seed vigor when soaked in distilled water. Conversely, soaking with "hard" water appeared to optimize vitamin C production with or without sunlight as was discussed earlier. Also, as discussed earlier, low levels of mineral salts, particularly Mn, were found to greatly enhance vitamin C production in various seed sprouts. Other well known promoters of seed germination are gibberellin salts, sulphydryl salts, nitrates and nitrites[305,306] as well as hydroxylamine and ammonium salts.[307] Finally, some sugars promoted vitamin C synthesis. In addition, as discussed in the section on minerals, Ganesh Kumar et al[124] reported 72 h germination resulted in losses of about 40% of the Ca and Mg from green gram and cowpea and about 20% of those minerals from chickpea, mostly due to leaching during steeping.

In addition to relatively large amounts of minerals and other water soluble seed components lost during distilled water soaking, an additional "loss" occurs since in distilled water there are no mineral salts or other water soluble nutrients or growth promoters present _to be absorbed_ by imbibing seeds. For example, as discussed earlier, researchers found Ca content of the sprouts naturally increased because of its presence in the water.[62,75,137] In addition Fordham et al[34] reported that Ca increases in sprouted soybeans were directly related to the concentration of chlorinated lime in the steep water.

Effects of Chlorine, Rinsing and Aeration on Sprouts

These effects are not covered in this chapter in depth but some remarks are offered in view of the importance of this topic. It is well known to commercial and home sprouters that aeration is required for effective sprouting but few studies include details about how this is accomplished and scarcely any provide quantitative data in this regard. In commercial barley malting operations it is a routine practice to continuously incorporate air into steep tanks, to move water saturated air through tumblers, or to periodically turn the germinating grain to insure oxygen availability and to exhaust accumulated CO_2. Multiple steeping to insure rapid germination should be included here.[308]

One recent study illustrates the effects of chlorine during rinsing and aeration on sprout quality. Hsu et al[37] added chlorine to tap water in the form of sodium hypochlorite, in concentrations of 50, 100, or 200 ppm, to study its effect on percent germination (4 days) and sprout development of yellow peas. There was no significant difference between 0 and 50 ppm chlorine on hypocotyl length. Above 50 ppm, chlorine decreased average hypocotyl length from 5.72 to 4.43 cm (Table 18). Higher concentrations of chlorine depressed average epicotyl length from 3.30 to 2.48 cm. It is particularly interesting that intermediate chlorine concentrations of 50 and 100 ppm significantly increased percent germination above that of tap water and 200 ppm chlorine treatment.

Aeration is an important component of steeping and germination of seeds. Sufficient oxygen must be supplied[309] and excess CO_2 exhausted. Far less well understood or publicized are the conditions which properly aerate during

Table 18. Effect of rinsing and chlorine on percent germination and sprout development of yellow pea.[37]

Treatment		Germination %	Hyplcotyl Length (cm)	Epicotyl Length (cm)
Manual rinsing	Tap water	83.7	5.8 ± 1.1	3.3 ± 0.8
	50 ppm chlorine	88.1	5.7 ± 1.6	2.7 ± 0.4
	100 ppm chlorine	86.0	5.0 ± 1.4	2.5 ± 0.4
	200 ppm chlorine	83.3	4.4 ± 1.5	2.5 ± 0.5
Automated rinsing	200 ppm chlorine soak and tap rinse	94.0	6.3 ± 1.6	3.0 ± 0.5
	Tap rinse & soak	92.0	6.4 ± 1.7	3.0 ± 0.5

germination. Wu and Fenton[33] suggested that the germination chamber should be thoroughly covered with wet cheese cloth "to prevent excess ventilation" which they believed was unfavorable to sprouting. Seeds respond favorably to an optimum amount of oxygen[280] and too much is as detrimental as too little. However, "excess ventilation" is less likely a problem except when air being moved over the seeds is too dry and tends to dehydrate them. In fact Hsu et al[37] found that vigorous, water saturated air movement greatly enhanced seed growth. Although preliminary in nature, that study showed that some air movement within the germination chambers is needed to ensure the most vigorous sprouting.

Automatic rinsing (every 2 h throughout the entire germination period) with tap water or 200 ppm chlorine steep was more effective than manual rinsing (Table 18).

TODAY'S FOOD PROBLEMS--TOMORROW'S FOOD NEEDS

Today, in spite of adequate food production, thousands of humans literally starve to death, and millions suffer from malnutrition caused by inadequate food or improperly balanced diets. Tomorrow, that tragic problem will become more severe as the world's population increases and more agricultural land is taken out of production. Forseably, increases in food production will diminish or cease, creating malnutrition or famine on an unprecedented scale.

One obvious way to help aleviate today's tragic problem of malnutrition and avoid possible catastrophy tomorrow is to stretch food supplies by improving the nutritive value of our foodstuffs, cereals and legumes. There are four basic ways to improve the nutritive quality of food and feed grains: use of more subtle and careful heat treatment, introduction of more precise nutrient balancing, expanded use of microbiological fermentation, and application of controlled germination.

Two outstanding nutritional biochemists summarized the major causes of disease and malnutrition among the subcontinental Indian population in a short paragraph followed by another one sentence paragraph, within which was offered a solution to those problems: "The major deficiencies in the diet are food energy and protein with associated deficiencies of calcium, vitamin A, riboflavin and other nutrients. The deficits in the former two can be met by increasing the proportion of dal (legume) and the total amount of food consumed. In the case of infants, pregnant women and the like of whom appetite, palatability and/or digestibility are limiting factors, food intake can be improved by the use of procedures such as sprouting, parching, roasting and fermentation which are commonly used in India. Those procedures can be further exploited for use in infant feeding. Such procedures not only help increase food intake but also increase the nutritive value of the foods." In a one sentence paragraph the author emphasized that: "A combination of those procedures e.g. roasting after soaking or sprouting and fermentation after sprouting is <u>even more effective</u> in attaining those objectives."[281]

Devadas and Murthy[282] realized no person, no family, "no nation can afford a generation of men and women incapable of achieving their full potentials of growth and working capacity. Therefore, it is imperative to launch preventative measures against malnutrition. Proper use of available indigenous nutrient rich food resources is an important step in combating malnutrition. Successful treatment of kwashiorkor with diets, based upon vegetable protein mixtures, especially legumes and oilseed meals, indicate that appropriate infant feeding and child feeding practices and use of local low cost vegetable foods will go a long way in controlling the problem."

"The type, quantities and proportion of the ingredients of low cost mixtures should be such as to bring about maximum supplementation in calories, proteins and other nutrients. How to give the vulnerable groups the additional calories and nutrients they need? One solution appears to be locating and using low-cost, indigenous, nutrient rich foods since foods of animal origin, such as milk, meat and egg are expensive and beyond the reach of most families. The urgency for the development of nutritionally balanced food products which would be within the reach of large segments of the population is critical."

By 1945, Burkholder and McVeigh[72] at the Osborn Botanical Laboratory, Yale University, had accumulated a sizable amount of data on nutritive benefits of germinating seeds. The authors wrote: "The significance of such data for human and animal nutrition need hardly be stressed. Our investigations indicate that marked increases of important nutrient substances occur during the sprouting of leguminous and many other kinds of edible seeds . . . The possible use of sprouted peas, beans, and other seeds in canning and preparation of dehydrated foods would enable these industries to operate on a 12 month basis. If the food value of germinated seeds is to be judged by their content of vitamins and readily available amino acids, then it appears that the common use of sprouts in the diets of oriental people rests on a sound nutritional basis and should be introduced on a wide scale among occidentals. It is hoped that the data reported here may have significance in academic studies on growth processes of plants and practical value in connection with the use of sprouted seeds for preparing high quality processed foods."

In summary, based (1) on nearly 100 years of chemical studies, (2) on about 70 years of corroborative rat and other animal feeding studies, (3) on further corroboration by a few well documented human feeding studies, and (4) on hundreds and in some cases thousands of years of experience by millions of people, it is concluded that carefully controlled, optimal germination of edible cereals and legumes is capable of significantly alleviating today's food problems and avoiding tomorrow's food needs.

ACKNOWLEDGMENTS

Thanks to Frank Loewus, Barry Swanson, Henry Leung, David Beguin and Renée Birch for reading the manuscript and for their helpful comments and suggestions. Thanks to Yasuto Sasaki for his translations of articles written in Japanese and to Patsy Allen for stenographic help.

REFERENCES

1. Furst, V. 1912 Weitere Beiträge zur Ätiologie des experimentellen Skorbuts des Meerschweinchens. Zeitschrift f. Hygiene LXXII:121-54.
2. Smith, A.H. 1918 Beer and Scurvy. Some notes from history. The Lancet 2:813-15.
3. Chick, H. and Hume, M. 1917 The distribution among foodstuffs (especially those suitable for the rationing of armies) of the substances required for the prevention of (A)Berberi and (B) scurvy. Transaction of the Society of Tropical Medicine and Hygiene. X:141-86.
4. Greig, E.D.W. 1917 The 'sprouting capacity' of grains issued as rations to troops. Indian J. Med. Res. 4:818-23.
5. Chick, H., Hume, E.M,, Skelton, R.F. and Smith, A.H. 1918 The relative content of antiscorbutic principle in limes and lemons together with some new facts and some old observations concerning the value of "lime juice" in the prevention of scurvy. The Lancet Nov. 30, 735-38.
6. Chick, H. and Delf, E.M. 1919 The anti-scorbutic value of dry and germinated seeds. Biochem. J. 13:199-218.
7. Wiltshire, H.W. 1918 A note on the value of germinated beans in the treatment of scurvy. The Lancet 2:811-13.
8. Dyke, H.W. 1918 Outbreak of scurvy in the South African Native Labour Corps. The Lancet. Oct. 19:513-15.
9. Harden, A. and Zilva, S.S. 1924 Investigation of barley, malt and beer for vitamins B and C. Biochem. J. 18:1129-32.
10. Delf, E.M. 1922 Studies in esperimental scurvy, with special reference to the antiscorbutic properties of some South African food stuffs. The Lancet. March 25:576-79.

11. Santos, F.O. 1922 Some plant sources of vitamin B and C. Am. J. Physiol. 59:310-34.

12. Honeywell, E.M. and Steenbock, H. 1924 The synthesis of vitamin C by germination. Am. J. Physiol. 70:322-32.

13. French, C.E., Berryman, G.H., Goorley, J.T., Harper, H.A., Harkness, D.M. and Thacker, E.J. 1944 The production of vitamins in germinated peas, soybeans, and other beans. J. Nutr. 28:63-70.

14. Wu, H. 1928 Nutritive value of Chinese foods. Chinese J. Physiol. 1:153-86.

15. Miller, C.D. and Hair, D.B. 1928 The vitamin content of mung bean sprouts. J. Home Ec. 20:263-71.

16. Wats, R.C. and Eyles, C.M.E. 1932 Some sources of vitamin C in India. Part II. Germinated pulses, tomatoes, mangoes and bananas. Ind. J. Med. Res. XX:89-106.

17. Bogart, R. and Hughes, J.S. 1935 Ascorbic acid (Vitamin C) in sprouted oats. J. Nutr. 10:157-60.

18. Ghosh, A.R. and Guha, B.C. 1935 Vitamin C in Indian Food Stuffs. J. Ind. Chem. Soc. 12:30-36.

19. Khan, M.M. 1942 Scurvy in the famine areas of Hissar district, Punjab. Ind. Med. Gaz. 72:6-14.

20. Johnson, S.W. 1933 The indophenol-reducing capacity and the vitamin C content of extracts of young germinated peas. Biochem. J. 27:1942-49.

21. Ray, S.N. 1934 On the nature of the precursor of the vitamin C in the vegetable kingdom. I. Vitamin C in the growing pea seedling. Biochem. J. 28:996-1003.

22. Ahmad, B. 1935 The vitamin C value of some common Indian fruits, vegetables and pulses by the chemical method. Ind. J. Med. Res. 22:789-99.

23. Ahmad, B. 1935 Observations on the chemical method for the estimation of vitamin C. Biochem. J. 29:275-81.

24. Lee, W.Y. 1936 The formation and distritution of vitamin C in germinating pea, <u>Pisum sativum L.</u> (Blue Bantam Variety). J. Chin. Chem. Soc. 4:219-23.

25. Harris, L.J. and Ray, S.N. 1933 Specificity of hexuronic (ascorbic) acid as antiscorbutic factor. Biochem. J. 27:580-89.

26. Glick, D. 1937 The quantitative distribution of ascorbic acid in the developing barley embryo. Compt. rend. Lab. Carlsberg, Serie chimique 21:203-9.

27. Biswas, H.G. and Das, K.L. 1937 A comparative study of vitamin C in a few germinated oilseeds. Science and Culture 3:176-77.

28. Chu, T.J. and Read, B.E. 1938 The vitamin C content of Chinese Foods. Part II. Chinese J. Physiol. 13:247-56.

29. Rudra, M.N. 1938 Studies in vitamin C. Part V. Vitamin C content of some germinated cereals and pulses. J. Indian. Chem. Soc. 15:191-93.

30. Muthanna, M.C. and Ahmad, B. 1940 Vitamin C content of germinated cereals. Current Science 7:320-21.

31. Bhagvat, K. and Narasinga Rao, K.K.P. 1942 Vitamin C in germinating grains. Ind. J. Med. Res. 30:493-504.

32. Nandi, N. and Banerjee, S. 1949 Studies on germination. Part II. The effect of germination on the vitamin C content of pulses grown in Bengal. Ind. Pharm. 5:63-68.

33. Wu, C.H. and Fenton, F. 1953 Effect of sprouting and cooking of soybeans on palatability, lysine, tryptophane, thiamine, and ascrobic acid. Food Research 18:640-45.

34. Fordham, J.R., Wells, C.E. and Chen, L.H. 1975 Sprouting of seeds and nutrient composition of seeds and sprouts. J. fd. Sci. 40:552-56.

35. Venugopal, K. and Rama Rao, G. 1978 Available lysine and vitamin C in germinated black gram (<u>Phaseolus mungo</u>) seeds. Ind. J. Nutr. Dietet. 15:9-11.

36. Bau, H.M. and Debry, G. 1979 Germinated soybean protein products: Chemical and nutritional evaluations. J. Am. Chem. Soc. 56:160-62.

37. Hsu, D., Leung, H.K., Finney, P.L., and Morad, M.M. 1980 Effect of germination on nutritive value and baking properties of dry peas, lentils, and faba beans. J. Fd. Sci. 45:87-92.

38. Bhagvat, K. and Narasinga Rao, K.K.P. 1942 Vitamin C content of dry bengal gram (<u>Cicer arietinum</u>). Ind. J. Med. Res. 30:505-11.

39. Hamilton, M.J. and Vanderstoep, J. 1979 Germination and nutrient composition of alfalfa seeds. J. Fd. Sci. 44:443-45.

40. Heller, V.G. 1927 Vitamin synthesis in plants as affected by light source. J. Biol. Chem. 76:499-511.

41. Lee, W.V. and Read, B.E. 1936 The effect of light on the production and distribution of ascorbic acid in germinating soybeans. J. Chin. Chem. Soc. 4:208-18.

42. Clark, W.G. 1937 Ascorbic acid in the avena coleoptile. Bot. Gaz. 99:116-24.

43. Reid, M.E. 1938 The effect of light on the accumulation of ascorbic acid in young cowpea plants. Am. J. Bot. 25:701-10.

44. Fa, W.C. and Roy, S. 1944 Affect of light on the synthesis of ascorbic acid by germinating seeds. Science & Culture 9:564.

45. Ahmad, B., Qureshi, A.A., Babbar, I. and Sawhney, P.C. 1946 Observations on ascorbic acid. Part IV. The effect of certain factors on ascorbic acid production during germination of seeds. Annals of Biochemistry and Experimental Medicine, Vol. VI:29-34, Calcutta, India.

46. Roy, S., Bose, A., and Guha, B.C. 1944 Photobiosynthesis of ascorbic acid (?) by germinating seeds. Science and Culture 9:564.

47. De, H.N. and Barai, S.C. 1949 Study of the mechanism of biosynthesis of ascorbic acid during germination. Ind. J. Med. Res. 37:101-11.

48. Prudente, V.R. and Mabesa, L.B. 1981 Vitamin content of mung bean [Vigna radiata (L.) R Wilcz.] sprouts. Phil. Agr. (U.P. Los Baños) 64:365-70.

49. Rudra, M.N. 1938 Role of manganese in the biological synthesis of ascorbic acid. Nature 142:203.

50. Rudra, M.N. 1939 Role of manganese in the biological synthesis of ascorbic acid. Nature 143:811.

51. Rudra, M.N. 1939 Role of manganese in the biological synthesis of ascorbic acid. Nature 144:868.

52. Hester, J.B. 1941 Manganese and vitamin C. Science 93:401.

53. Guha, B.C. and Ghosh, A.R. 1934 Synthesis of ascorbic acid (Vitamin C) by means of tissues in vitro. Nature 134:739.

54. Guha, B.C. and Ghosh, A.R. 1935 Biological formation of ascorbic acid. Nature 135:234.

55. Guha, B.C. and Ghosh, A.R. 1935 Biological synthesis of Vitamin C. Nature 135:871.

56. Guha, B.C. and Ghosh, A.R. 1936 Biosynthesis of ascorbic acid. Nature 136.

57. Virtanen, A.I. and Eerola, L.V. 1936 Formation of vitamin C in germinating seeds. Suomen Kemistilehti B NO. 4:13-16.

58. Havas, L. 1935 Ascorbic acid (Vitamin C) and the germination and growth of seedlings. Nature 135:435.

59. von Hausen, S. 1935 Effect of vitamin C (Ascorbic acid) on the growth of plants. Nature 135:516.

60. von Hausen, S. 1935 Effect of vitamin C (Ascorbic acid) on the growth of plants. Suomen Kemistilehti B NO. 5:27-28.

61. Davies, W., Atkins, G.A., and Hudson, P.C.B. 1937 The affect of ascorbic acid and certain indole derivatives on the regeneration and germination of plants. Annals. Bot. 1:329-51.

62. Embrey, H. 1921 The investigation of some Chinese foods. The China Med. J. 35:420-47.

63. Bowman, H.H.M. and Yec, M.A. 1925 Crystals of vitamin B from the mung bean. Proc. Soc. Exp. Biol. & Med. xxii:228-31.

64. Rose, M.S. and Phipaid, E.H.F. 1937 Vitamin B and G values of peas and lima beans under various conditions. J. Nutr. 14:55-67.

65. Swaminathan, M. 1942 An improved method for the estimation of vitamin B_1 in foods by the thiochrome reaction. Ind. J. Med. Res. 30:263-72.

66. Burkholder, P.R. 1943 Vitamins in edible soybeans. Science 98:188-90.

67. Daniel, L. and Norris, L.C. 1945 The riboflavin, niacin and thiamine content of dried leguminous seeds. J. Nutr. 30:31-36.

68. Stamberg, O.E. and Lehrer, W.P. 1947 Composition, including thiamine and fiboflavin, of edible dry legumes. Fd. Res. 12:270-72.

69. Burkholder, P.R. and McVeigh, I. 1942 The increase of B vitamins in germinating seeds. Proc. Nat. Acad. Sci. 28:440-46.

70. Cheldelin, V.H. and Lane, R.L. 1943 B vitamins in germinating seeds. Proc. Soc. Biol. Med. 54:53-55.

71. Burkholder, P.R. 1943 Vitamins in dehydrated seeds and sprouts. Science 97:562-64.

72. Burkholder, P.R. and McVeigh, I. 1945 Vitamin content of some mature and germinated legume seeds. Plant Physiol. 20:301-5.

73. Wai. K.N.T., Bishop, J.C., Mack, P.B. and Cotton, R.H. 1947 The vitamin content of soybeans and soybean sprouts as a function of germination time. Plant Physiol. 22:117-26.

74. Chattopadhyay, H., Nandi, N., and Banerjee, S. 1950 Studies on germination. Part III. The effect of germination on the thiamine content of the pulses grown in Bengal. Ind. Pharm. 5:121-22.

75. Kylen, A.M. and McCready, R.M. 1975 Nutrients in seeds and sprouts of alfalfa, lentils, mung beans and soybeans. J. Food Sci. 40:1008-9.

76. Davis, C.F., Laufner, S. and Saletan, L. 1943 A study of some of the vitamin B complex factors in malted and unmalted barley and wheat of the 1941 crop. Cereal Chem. 20:109-13.

77. McVeigh, I. 1944 Occurrence and distribution of thiamine, riboflavin, and niacin in avena seedlings. Bulletin of the Torrey Botanical Club. 71:438-44.

78. Chen, L.H., Wells, C.E. and Fordham, J.R. 1975 Germinated seeds for human consumption. J. Food. Sci. 40:1290-94.

79. Lee, F.A. and Whitcomb, J. 1945 Effect of freezing preservation and cooking on vitamin content of green soybeans and soybean sprouts. J. Am. Dietet. Assoc. 21:696-97.

80. Klatzkin, C., Norris, F.W. and Wokes, F. 1945 Riboflavin in malt extract. Quart. J. Pharm. and Pharmocology. 19:376-87.

81. Hopkins, R.H. and Wiener, S. 1944 Vitamin B_1 and riboflavin in brewing. J. Inst. Brew. 50:124-38.

82. Hopkins, R.H. 1945 Vitamins in top fermentation brewing materials and beer. Wallerstein Comm. 8:110-17.

83. Stringer, W.J.1946 II. Vitamins in beer. J. Inst. Brew. 52:81-87.

84. Knorr, F. 1952 Vitamine in der brauerei. Brauwissenschaft. Heft. 4:70-71.

85. Elvehjem, C.A., Madden, R.J., Strong, F.M. and Wolley, D.W. 1937 Relation of nicotinic acid and nicotinic acid amide to canine black tongue. J. Am. Chem. Soc. 59:1767-68.

86. Mason, J.B., Gibson, N., and Kodicek, E. 1973 The chemical nature of the bound nicotinic acid of wheat bran: Studies of nicotinic acid-containing macromolecules. Br. J. Nutr. 30:297-311.

87. Kodicek, E. 1940 Estimation of nicotinic acid in animal tissues, blood and certain food stuffs. I. Method. Biochem. J. 34:712-23.

88. Kodicek, E. 1940 Estimation of nicotinic acid in animal tissues, blood and certain food stuffs. II. Applications. Biochem. J. 34:724-35.

89. Swaminathan, M. 1944 Nicotinic acid content of Indian Food stuffs. Ind. J. Med. Res. 32:39-46.

90. Klatzkin, C., Norris, F.W. and Wokes, F. 1948 Nicotinic acid in Cereals. I. The effect of germination. Biochem. J. 42:414-20.

91. Nandi, N., and Banerjee, S. 1949a Studies on germination. Part 1. The effect of germination on the nicotinic acid content of pulses grown in Bengal. Ind. Pharm. 5:13-16.

92. Banerjee, S., Rohatgi, K. and Lahiri, S. 1954 Pantothenic acid. folic acid, biotin, and niacin contents of germinated pulses. Food Research 19:134-37.

93. Swaminathan, M. 1940 A chemical test for vitamin B_6 in foods. Ind. J. Med. Res. 28:427-39.

94. Sarma, P.S. 1944 The estimation of pyridoxine (vitamin B_6) in foods using recemoth larvae (Corcyra cephalonica St.). Ind. J. Med. Res. 32:117-22.

95. Esh, G.C. 1955 Studies on the nutritive value of plant proteins. Part II. Influence of vitamin B_{12} the nutritive value of pulse proteins. Ind. J. Physiol. and All. Sci. IX:129-33.

96. Rohatgi, K., Banerjee, M., and Banerjee, S. 1955 Effect of germination on vitamin B_{12} values of pulses (leguminous seeds). J. Nutr. 56:403-8.

97. Ramachandran, M. and Phansalkar, S.V. 1956 Absence of true vitamin B_{12} activity of pulses. Current Science. 25:260.

98. Lavate, W.V. and Sreenivasan, A. 1956 Absence of vitamin B_{12} in sprouted mung (*Phaseolus radiatus*). J. Sci. Industr. Res. 15c:213-14.

99. Hofsten, Bengt 1979 Legume sprouts as a source of protein and other nutrients. J. M. Oil Chem. Soc. 56:382.

100. Hofsten 1978 Personal communication by letter.

101. Lehninger, A.L. 1975 Biochemistry: the molecular basis of cell structure and function, 2nd Edition, Worth Publishers, N.Y., N.Y.

102. Chattopadhyay, H. and Banerjee, S. 1951 Effect of germination on the carotene content of pulses and cereals. Science 113:600-1.

103. Chattopadhyay, H. and Banerjee, S. 1952 Effect of germination on the total tocopherol content of pulses and cereals. Food Research 17:402-3.

104. Dam, H., Glavind, J. and Svendsen, I. 1938 Vitamin K in the plant. Biochem. J. 32:485-87.

105. Erkama, J. and Pettersson,N. 1950 Vitamin K in germinating Peas. Acta Chemica Scandinavica 4:922-25.

106. Banerjee, S., Rohatgi, K., Banerjee, M., Chattopadhyay, D. and Chattopadhyay, H. 1955 Pyridoxine, Inositol, and vitamin K contents of germinated pulses. Food Res. 20:545-47.

107. Peers, R.G. 1953 The phytase of wheat. Biochem. J. 53:102-110.

108. Belavady, B., and Banerjee, S. 1952 Studies on the effect of germination on the phosphorus values of some common Indian pulses. Food Res. 18:223-26.

109. Darbre, A. and Norris, F.W. 1956 Vitamins in germination. Determination of free and combined inositol in germinating oats. Biochem. J. 64:441-46.

110. Darbry, A. and Norris, F.W. 1957 Determination of free and combined inositol in the ungerminated bean seed and the yound plant. Biochem. J. 66:404-7.

111. Mayer, A.M. 1958 The breakdown of phytin and phytase activity in germinating lettuce seeds. Enzymologia 19:1-8.

112. Ergle, D.R. and Guinn, G. 1959 Phosphorous compounds of cotton embryos and their changes during germination. Plant Physiol. 34:476-81.

113. Sobolev, A.M. 1962 Enzymatic hydrolysis of phytin *in vitro* and in germinating seeds. Soviet Plant Physiol. 9:263-69.

114. Gibbins, L.N. and Norris F.W. 1963 Vitamins in germination. Distribution of inositol during the germination of the dwarf beans *Phaseolus vulgaris*. Biochem. J. 86:64-67.

115. Richardson, K.E. and Axelrod, B. 1957 Changes in the inositol content during germination and growth of some higher plants. Plant Physiol. 32:334-37.

116. Milailovic, M.L., Antic, N. and Hadzijev, D. 1965 Chemical investigation of wheat. 8. Dynamics of various forms of phosphorous in wheat during its ontogenesis. The extent and mechanism of phytic acid decomposition in germinating wheat grain. Plant Soil 23:117-28.

117. Tomlinson, R.V. and Ballou, C.E. 1962 Myoinositol polyphosphate intermediates in the dephosphorylation of phytic acid by phytase. Biochem. 1:166-73.

118. Shrivastava, B.I. 1964 The effect of GA on ribonuclease and phytase activity of germinating barley seeds. Canadian J. Bot. 42:1303-11.

119. Matheson, N.K. and Strother, S. 1969 The utilization of phytate by germinating wheat. Phytochem. 8:1349-56.

120. Bianchetti, R. and Sartirana, M.L. 1967 The mechanism of the repression by inorganic phosphate of phytase synthesis in the germinating wheat embryo. Biochimica et Biophysica Acta. 145:485-93.

121. Mandal, N.C. and Biswas, B.B. 1970 Metabolism of inositol phosphates. I. Phytase synthesis during germination in cotyledons of mung bean, *Phaseolus aureus*. Plant Physiol. 45:4-7.

122. Chen, L.H. and Pan, S.H. 1977 Decrease of phytates during germination of pea seeds (*Pisum sativa*). Nutr. Reports Inter. 16:125-30.

123. Reddy, N.R., Balakrishnan, C.V. and Salunkhe, D.K. 1978 Phytate phosphorus and mineral changes during germination and cooking of black gram (Phaseolus mungo) seeds. J. Fd. Sci. 43:540-43.

124. Ganesh Kumar, K., Venkataraman, L.V., Jaya, T.V. and Kirshnamurthy, K.S. 1978 Cooking characteristics of some germinated legumes: changes in phytins, Ca++, Mg++, and pectins. J. Fd. Sci. 43:85-88.

125. Mandal, N.C., Burman, S. and Biswas, B.B. 1972 Isolation, purification and characterization of phytase from germinating mung beans. Phytochem. 11:495-502.

126. Kon, S., Olson, A.C., Frederick, D.P., Eggling, S.B. and Wagner, J.R. 1973 Effect of different treatments on phytate and soluble sugars in California small white beans. J. Fd. Sci. 38:215-17.

127. Chang, R., Schwimmer, S. and Burr, H.K. 1977 Phytate:removal from whole dry beans by enzymatic hydrolysis and diffusion. J. Fd. Sci. 42:1098-1101.

128. Tabekhia, M.M. and Luh, B.S. 1980 Effect of germination, cooking, and canning on phosphorous and phytate retention in dry beans. J. Fd. Sci. 45:406-408.

129. Finney, P.L., Mason, W.R., Jeffers, H.C., El-Samahy, S.K. and Vigue, G.T. 1981 Effects of germination on some physical, chemical, and breadmaking properties of 12 U.S. wheat variety composites. Paper #214. 66th Annual meeting of the AACC, Denver, CO, Oct. 25-29, 1981.

130. Chattopadhyay, H. and Banerjee, S. 1951 Studies on the choline content of some common Indian pulses and cereals both before and during the course of germination. Food Res. 16:230-32.

131. Cravioto, R., Lockhart, E.E., Anderson, R.K., Miranda, F.P. and Harris, R.S. 1945 Composition of typical Mexican foods. J. Nutr. 29:317-329.

132. Patwardhan, V.N. 1962 Pulses and beans in human nutrition. Am. J. Clin. Nutr. 11:12-30.

133. Cain, R.F. 1977 Final report project No. NWI-09 Nutritional evaluation of varieties and classes of wheat. Part I. Proximate analysis, amino acids, thiamine, riboflavin, niacin, pyridoxine and minerals, a cooperative project (3 parts) among Univeristy of Idaho, Oregon State University, and Washington State University. Mimeograph Report.

134. Gopalan, C., Ramasastri, B.V. and Balasubramanian, S.C. 1971 Nutritive value of Indian Foods. National Institute of Nutrition. ICMR, Hyderabad.

135. Singh, H.D. and Banerjee, S. 1955 Studies of the effect of germination on the availability of iron in some common Indian pulses. Ind. J. Med. Rs. 43:497-500.

136. Rajalakshmi, R. and Patel, I. 1969 Unpublished, but noted in : Rajalakshmi, R. (1969) Applied Nutrition. 1st Edition. Oxford and IBH Publishing, New Delhi.

137. Beal, L., Mehta, T. and Finney, P.L. 1982 Germination improves Zn availability from peas. Paper #467, 42nd Annual meeting of IFT, Las Vegas, June 22-25, 1982.

138. Albaum, H.G. and Eichel, B. 1943 The relationship between growth and metabolism in the oat seedling. Am. J. Bot. 30:18-22.

139. Kornberg, H.L. and Beevers, H. 1957 The glyoxylate cycle as a stage in the conversion of fat to carbohydrate in castor beans. Biochimica et Biophysica. acta:26:531-37.

140. Canvin, D.T. and Beevers, H. 1960 Sucrose synthesis from acetate in the germinating caster bean: Kinetics and pathway. J. Bio. Chem. 236:988-95.

141. Yound, J.L. and Varner, J.E. 1959 Enzyme synthesis in the cotyledons of germinating seeds. Arch. Biochem. Biophy. 84:71-8.

142. Spragg, S.P. and Yemm, E.W. 1959 Respiratory mechanisms and the changes of glutathione and ascorbic acid in germinating peas. J. Exptl. Bot. 10:409-25.

143. Brown, A.P. and Wray, J.L. 1968 Correlated changes of some enzyme activities and cofactor and substrate contents of pea cotyledon tissue during germination. Biochem. J. 108:437-44.

144. Kollöffel, C. 1968 Activity of alcohol dehydrogenase in the cotyledons of peas germinated under different environmental conditions. Act. Bot. Nerrl. 17:70-7.

145. Fernandez-Terrago, J., Rodrigues-Bujan, M.C. and Nicolas, G. 1978 Starch degradation during germination of Cicer arietinum L seeds. Rev. esp. Fisiol. 43:87-91.

146. Finney, P.L., Beguin, D. and Hubbard, J.D. 1982 Some effects of germination on the bread baking properties of mung bean (Cicer arietinum). Cereal Chem. IN PRESS.

147. De La Fuente Burguillo, P. and Nicolas, G. 1974 Respiratory activity during germination of seeds of Cicer arietinum L. I. Glycolysis and fermentation. Plant science letters 3:143-48.

148. Nigam, V.N. and Giri, K.V. 1961 Sugar in pulses. Can J. Biochem. & Physiol. 39:1847-53.

149. Pazur, J.H., Shadaksharaswamy, M. and Meidell, G.E. 1962 The metabolism of oligosaccharides in germinating soybeans, Glycine max. Archives of Biochemistry and Biophysics 99:78-85.

150. Subbulakshmi, G., Ganesh Kumar, K. and Venkataraman, L.V. 1976 Effect of germination on the carbohydrates, proteins, trypsin inhibitor, amylase inhibitor and hemagglutinin in horsegram and moth bean. Nutr. Reports Inter. 13:19-31.

151. Ganesh Kumar, K. and Venkataraman, L.V. 1976 Studies on the in vitro digestibility of starch in some legumes before and after germination. Nutr. Rep. Inter. 13:115-24.

152. Warsi, S.A., Fatima, R. and Qadri, R.B. 1977 Effect of soaking and germination on the nutritive value of gram and pea. Pak. J. Biochem. 10:55-64.

153. Åman, P. 1979 Carbohydrates in raw and germinated seeds from mung bean and chickpea. Sci. Fd. Agric. 30:869-75.

154. Silva, H.C. and Luh, B.S. 1979 Changes in oligosaccharides and starch granules in germinating beans. Can. Inst. Fd. Sci. Technol. J. 12:103-7.

155. Gupta, K. and Wagle, D.S. 1980 A research note: Changes in antinutritional factors during germination in Phaseolus mungoreous, a cross between Phaseolus mungo (M_1-1) and Phaseolus aureus (T_1). J. Fd. Sci. 45:394-95, 397.

156. Azhur, S., Srivastava, A.k. and Krishna Murthy, C.R. 1972 Compositional changes during the germination of Cicer arietinum. Phytochem. 11:3173-79.

157. Calloway, D.H., Hickey, C.A. and Murphy, E.L. 1971 Reduction of intestinal gas-forming properties of legumes by traditional and experimental food processing methods. J. Food Sci. 36:251-55.

158. Narayana Rao, M., Shurpalekar, K.S., Sundaravalli, E.E. and Doraiswamy, T.R. 1973 Flatus production in children fed legume diets. Pag. Bulletin, Vol. III, No.2.

159. Shurpalekar, K.S., Sundaravalli, O.E. Desai, B.L.M. 1973 Effect of cooking and germination on the flatus inducing capacity of some legumes. In: Nutritional aspects of common beans and other legume seeds as animal and human feed. Ed. Jagge, W.G. Archievos Latinamericanos de Nutricion:133-37.

160. Rao, P.S. 1969 Studies on the digestibility of carbohydrates in pulses. Ind. J. Med. Res. 57:2151-57.

161. Venkataraman, L.V. and Jaya, T.V. 1975 Gastrointestinal gas production in rats fed on diets containing germinated legumes. Nutr. Reports Inter. 12:387-98.

162. Rao, P.U. and Belavady, B. 1978 Oligosaccharides in pulses: Varietal differences and effects of cooking and germination. J. Agric. Fd. Chem. 26:316-19.

163. Jaya, L.V., Naik, H.S. and Venkataraman, L.V. 1979 Effect of germinated legumes on the rate of in vitro gas production by Clostridium perfringens. Nutr. Rep. Inter. 20:393-401.

164. Talwalker, R.T., Garg, N.K. and Krishna Murti, C.R. 1964 Lipid changes in germinating Gram (Cicer arietinum). Ind. J. Exptl. Biol. 2:37-40.

165. Talwalker, R.T., Garg, N.K. and Krisha Murti, C.R. 1969 Phospholipase D (Phosphatidylcholine Phosphatidohydrolase, EC 3.1.4.4) of Cicer arietinum. Ind. J. Biochem. 6:228-30.

166. Lemar, L.E. and Swanson, B.G. 1976 Nutritive value of sprouted wheat flour: A research note. J. Fd. Sci. 41:719-20.

167. Ranhotra, G.S., Loewe, R.J. and Lehmann, T.A. 1977 Breadmaking quality and nutritive value of sprouted wheat. J. Fd. Sci. 42:1373-75.

168. Wu, Y.V. and Wall, J.S. 1980 Lysine content of protein increased by germination of normal and high-lysine sorghums. J. Agric. Fd. Chem. 28:455-58.

169. Beevers, L. 1968 Protein degradation and procolytic activity in the cotyledons of germinating pea seeds. (Pisum sativum). Phytochemistry 7:1837-44.

170. Smith, D.L. and Flinn, A.M. 1967 Histology and histochemistry of the cotyledons of Pisum arvense L. during germination. Planta (Berl.) 74:72-85.

171. Horner, H.T. and Arnott, H.J. 1965 A histochemical and ultrastructural study of yucca seed proteins. Amer. J. Bot. 52:1027-38.

172. Briarty, L.G., Coult, D.A. and Boulter, D. 1970 Protein bodies of germinating seeds of Vicia faba. Changes in fine structure and biochemistry. J. Exp. Botany. 21:513-24.

173. Tombs, M.P. 1967 Protein bodies in the soybean. Ph. Physiol., Lancaster 42:797-813.

174. Bagley, B.W., Cherry, J.H., Rollins, M.L. and Altschul , A.M. 1963 A study of protein bodies during germination of peanut (Arachis hypogea) seed. Am. J. Bot. 50:523-32.

175. Öpik, H. 1966 Changes in cell fine structure in the cotyledons of Phaseolus vulgaris L. during germination. J. Exp. Bot. 17:427-38.

176. Marcus, A. and Feeley, J. 1964 Activation of protein synthesis in the imbibition phase of seed germination. Proc. Natl. Acad. Sci. 51:1075-79.

177. Marcus, A. and Feeley, J. 1965 Protein synthesis in imbibed seeds. II. Polysome formation during imbibition. J. Biol. Chem. 240:1675-80.

178. Marcus A. and Feeley, J. 1966 Protein synthesis in imbibed seeds. III. Kinetics of amino acid incorporation, ribosome activation and polysome formation. Plant Physiol. 41:1167-72.

179. Yomo, H. and Taylor, M.P. 1973 Histochemical studies on protease formation in the cotyledons of germinating bean seeds. Planta (Berl.) 112:35-43.

180. Ganesh Kumar, K. and Venkataraman, L.V. 1978 Chickpea seed proteins: Modification during germination. Phytochemistry 17:605-9.

181. Cerletti, P., Fumagalli, A and Venturin, D. 1978 Protein composition of seeds of Lupinus albus. J. Fd. Sci. 43:1409-12.

182. Swain, R.R. and Dekker, E.E. 1966 Seed germination studies. I. Purification and properties of an α-amylase from the cotyledons of germinating peas. Biochem. Biophys. Acta 122:75-86.

183. Swain, R. and Dekker, E.E. 1966 Seed germination studies. II. Pathways for starch degradation in germinating pea seedlings. Biochem. Biophys. Acta 122:87-100.

184. Kumar, K.G. and Venkataraman, L.V. 1978 Chickpea seed proteins: Modification during germination. Phytochemistry 1:605-609.

185. Belton, W.E. and Hoover, C.A. 1948 Investigations on the Mung bean (Phaseolus aureus, Roxburgh) I. The determination of eighteen amino acids in the mung bean hydrolysate by chemical and microbiological methods. J. Biol. Chem. 175:377-83.

186. Bagchi, S.P., Ganguli, N.C. and Roy, S.C. 1955 Amino acid composition of some Indian pulses: quantitative determination by paper chromatography. Ann. Biochem. Exp. Med. 15:149-54.

187. Vijayaraghavan, P.K. and Srinivasan, P.R. 1953 Essential amino acid composition of some common Indian pulses. J. Nutr. 51:261-271.

188. Desikachar, H.S.R. and De, S.S. 1947 The cystine and methionine contents of common Indian Food-stuffs. Current Science 9:284.

189. Everson, G, and Heckert, A. 1944 The biological value of some leguminous sources of protein. J. Am. Diet. Assoc. 20:81-2.

190. Cook, A.H. 1962 Barley and Malt Biology, Biochemistry, Technology. Academic Press, New York, NY.

191. Tsai, C.Y., Dalby, A. and Jones, R.A. 1975 Lysine and tryptophan increases during germination of maize seed. Cereal Chem. 52:356-60.

192. Dalby, A. and Tsai, C.Y. 1976 Lysine and tryptophan increases during germination of cereal grains. Cereal Chem. 53:222-27.

193. Mason, W. and Finney, P.L. 1977 Information from M.S. thesis, Department of Agronomy and Genetics, Washington State University, Pullman, WA. Work completed in collaboration with USDA, ARS, Western Wheat Quality Laboratory, WSU, Pullman, WA.

194. Finney, P.L., Morad, M.M. and Hubbard, J.D. 1980 Germinated and ungerminated faba bean in conventional U.S. breads made with and without sugar and in Egyptian Balady breads. Cereal Chem. 57:267-70.

195. Johns, C.O. and Finks, A.J. 1920 Studies in nutrition. II. The role of cystine in nutrition as examplified by nutrition experiments with the proteins of the navy bean, Phaseolus vulgaris. J. Biol. Chem. 41:379-89.

196. Basu, K.P., Nath, M.C. and Ghani, M.O. 1936 Biological value of the proteins of green gram (Phaseolus mungo) and lentil (Lens esculenta) Part I. By the balance sheet method. Ind. J. Med. Res. 23:789-810.

197. Basu, K.P., Nath, M.C. and Ghani, M.O. 1936 Biological value of the proteins of green gram (Phaseolus mungo) and lentil (Lens esculenta) Part II. Measured by the growth of young rats. Ind. J. Med. Res. 23:811-26.

198. Russell, W.C., Taylor, M.W., Mehrhof, T.G. and Hirsh, R.R. 1946 The nutritive value of the protein of varieties of legumes and the effect of methionine supplementation. J. Nutr. 30:313-25.

199. Richardson, L.R. 1948 Southern peas and other legume seeds as a source of protein for the growth of rats. J. Nutr. 36:451-62.

200. Sherwood, F.W., Weldon, V. and Peterson, W.J. 1954 Effect of cooking and of methionine supplementation on the growth-promoting property of cowpea (Vigna sinensis) protein. J. Nutr. 52:199-208.

201. Schneider, B.H. and Miller, D.F. 1954 The biological value of alaska pea proteins. J. Nutr. 52:581-90.

202. Hirwe, R. and Magar, N.G. 1953 Effect of autoclaving on the nutritive value of pulses. Ind. J. Med. Res. 41:191-200.

203. Phansalkar, S.V., Ramachandran, M. and Patwardhan, V.N. 1957 Nutritive value of vegetable proteins. Part I. Protein efficiency ratio of cereals and pulses and the supplementary effect of the addition of a leafy vegetable. Ind. J. Med. Res. 45:611-21.

204. Phansalkar, S.V., Ramachandran, M. and Patwardhan, V.N. 1958 Nutritive value of vegetable proteins. Part II. The effect of vegetable protein diets on the regeneration of haemoglobin and plasma proteins in protein-depleated rats. Ind. J. Med. Res. 46:333-43.

205. Kuppuswamy, S., Srinivasan, M. and Subramanyan, V. 1958 Proteins in Foods. Indian Council of Medical Research, Wesley Press, Mysore City, Mysore, INDIA.

206. Kakade, M.L. and Evans, R.J. 1965 Nutritive value of navy beans (Phaseolus vulgaris). Brit. J. Nutr. 19:269-76.

207. Everson, G.J., Steenbock, H., Cedrequist, D.C. and Parsons, H.T. 1943 The effect of germination, the stage of maturity, and the variety upon the nutritive value of soybean protein. J. Nutr. 27:225-29.

208. Mattingly, J.P. and Bird, H.R. 1945 Effect of heating, under various conditions, and of sprouting on the nutritive value of soybean oil meals and soybeans. Poultry Sci. 24:344-52.

209. Chattopadhyay, H. and Banerjee, S. 1953 Effect of germination on the biological value of proteins and the trypsin-inhibitor activity of some common Indian pulses. Ind. J. Med. Res. 41:185-89.

210. Nadkarni, K.M. 1954 Indian materia medica, Vol I (Popular Book Depot, Bombay).

211. Grieve, M. 1959 A modern herbal. Hafner Pub. Co., New York, N.Y.

212. Rajalakshmi, R. and Subbulakshmi, G. 1964 Effect of fenugreek (Trigonella foenum-graeaim Linn.) supplementation on the biological value of rice and black gram (Phaseolus mungo) diet. Ind. J. Biochem. 1:104-6.

213. Jaya, T.V., Krishnamurthy, K.S. and Venkataraman, L.V. 1975 Effect of germination and cooking on the protein efficiency ratio of some legumes. Nutr. Rep. Inter. 12:175-82.

214. Bates, R.P., Knapp, F.W. and Araujo, P.E. 1977 A research Note: Protein quality of green-mature, dry-mature and sprouted soybeans. J. Fd. Sci. 42:271-72.

215. Khan, M.A. and Ghafoor, A. 1978 The effect of soaking, germination and cooking on the protein quality of mash beans (Phaseolus mungo). J. Sci. Fd. Agric. 29:461-64.

216. Jaya, T.V. and Venkataraman, L.V. 1979 Effect of germination on the supplementary value of chickpea and green gram protein to those of rice and wheat. Nutr. Rep. Inter. 19:777-83.

217. Bhagvat, K. and Sreenivasaya, M. 1935 The "Non-Protein" nitrogen of pulses. Current Sci. 3:354-55.

218. Swaminathan, M. 1938 The relative amounts of the protein and non-protein nitrogenous constituents occuring in foodstuffs and their significance in the determination of the digestibility co-efficient of proteins. Ind. J. Res. 25:847-55.

219. Swaminathan, M. 1939 The relative value of the proteins of certain foodstuffs in nutrition. Part V. Supplementary values of the proteins of Eleusine coracana (Ragi) and of certain pulses and skimmed milk powder studied by the nitrogen balance and the growth method. Ind. J. Med. Res. 26:107-12.

220. Venkataraman, L.V., Jaya, T.V. and Krishnamurthy, K.S. 1976 Effect of germination on the biological value, digestibility coefficient and net protein utilization of some legume proteins. Nutr. Rep. Inter. 13:197-204.

221. El-Hag, N., Haard, N.F. and Morse, R.E. 1978 Influence of sprouting on the digestibility coefficient, trypsin inhibitor and globulin proteins of red kidney beans. J. Fd. Sci.43:1874-75.

222. Subra Rao, P.V. and Desikachar, H.S.R. 1964 Indigestible residue in pulse diets. Ind. J. Expt. Biol. 2:243-44.

223. Borchers, R. and Ackerson, C.W. 1947 Trypsin inhibitor. IV. Occurence in seeds of the Leguminosae and other seeds. Arch. Biochem. 13:291-93.

224. Borchers, R. and Ackerson, C.W. 1950 The nutritive value of legume seeds. X. Effect of autoclaving and the Trypsin inhibitor test for 17 species. J. Nutr. 41:339-45.

225. Osborne, T.B. and Mendal, L.B. 1917 The use of soybeans as food. J. Biol. Chem. 32:369-79.

226. Jaffe, W.G. 1949 Toxicity of raw kidney beans. Experientia 5:81-87.

227. Heintze, K. 1950 Über Hitzebehandlung von Leguminosen und ihre Wirkung auf Geschack, proteolytische Inhibitoren und biologische Wertigkeit. Z. Lebensm. Untersuch. Forsch 91:100-111.

228. Phadke, K. and Sohone, K. 1962 Nutritive value of field bean (Dolichos lablab). II. Effect of feeding raw, autoclaved, and germinated beans on the growth of rats and nitrogen balance studies. J. Sci. Ind. Res. (India) 21:178-80.

229. Viswanatha, T. and De, S.S. 1951 Relative availability of cystine and methionine in the raw germinated and autoclaved soybeans and soybean milk. Ind. J. Physiol. Allied . Sci. 5:51-58.

230. Davadatta, S.C., Acharya, B.N. and Nadkarni, S.B. 1951 Effect of germination on the nutritional qualities of some of the vegetable proteins. Proc. Indian Acad. Sci. Sect. B33:150-58.

231. Kakade, M.L. and Evans, R.J. 1966 Growth inhibition of rats fed raw navy beans (Phaseolus vulgaris). J. Nutr. 90:191-98.

232. Palmer, R., McIntosh, A., and Pusztai, A. 1973 The nutritional evaluation of kidney bean (Phaseolus vulgaris). The effect on nutritional value of seed germination and changes in trypsin inhibitor content. J. Sci. Fd. Agric. 24:937-44.

233. El-Hag, N., Haard, N.F. and Morse, R.E. 1978 Influence of sprouting on the digestibility coefficient, trypsin inhibitor and globulin proteins of red kidney beans. J. Fd. Sci. 43:1874-75.

234. Gupta, K. and Wagle, D.S. 1980) A research note: Changes in antinutritional factors during germination in Phaseolus mungareous, a cross between Phaseolus mungo (M_1-1) and Phaseolus aureus (T_1). J. Fd. Sci. 45:394-95, 397.

235. Pusztai, A. 1972 Metabolism of trypsin-inhibitory proteins in the germinating seeds of kidney bean (Phaseolus vulgaris). Planta (Berl.) 107:121-29.

236. Pusztai, A. 1966 The isolation of two proteins. Glycoprotein I and a trypsin inhibitor from the seeds of kidney bean (Phaseolus vulgaris). Biochem. J. 101:379-86.

237. Wagner, L.P. and Riehm, J.P. 1967 Purification and partial characterization of a trypsin inhibitor from the navy bean. Arch. Biochem. Biophys. 121:672-81.

238. Pusztai, A. 1968 General properties of a protease inhibitor from the seeds of kidney bean. Eur. Biochem. 5:252-59.

239. Freed, R.C. and Ryan, D.S. 1978 Changes in Kunitz trypsin inhibitor during germination of soybeans: An immunoelectrophoresis assay system. J. Fd. Sci. 43:1316-19.

240. Kwong, E. and Barnes, R.H. 1963 Effect of soybean inhibitor on methionine and cystine utilization. J. Nutr. 81:392-98.

241. Hobday, S.M., Thurman, D.A. and Barber, D.J. 1973 Proteolytic and trypsin inhibitory activities in extracts of germinating Pisum sativum seeds. Phytochem. 12:1041-46.

242. Basu, K.P., Basak, M.N. and De, H.N. 1941 Studies in human nutrition. Part III. Protein, calcium, and phosphorus metabolism with typical Indian dietaries. Ind. J. Med. Res. 29:105-17.

243. Venkatachalam, P.S., Srikantia, S.G., Mehta, G. and Gopalan, C. 1956 Treatment of nutritional oedema syndrome (kwashiorkor) with vegetable diets. Ind. J. Med. Res. 44:539-45.

244. Venkatachalam, P.S., Srikantia, S.G. and Gopalan, C. 1954 Clinical features of nutritional oedemia syndrome in children. Ind. J. Med. Res. 42:555-68.

245. Rajalakshmi, R. and Ramakrishnan, C.V. 1977 Formulation and evaluation of meals based on locally available foods for young children. Wld. Rev. Nutr. Diet. 27:34-104.

246. Rajalakshmi, R. 1969 Applied nutrition, 1st Edition(Oxford and IBH Publishing, New Delhi).

247. Rajalakshmi, R., and Vanaja, K. 1967 Chemical and biological evaluation of the effects of fermentation on the nutritive value of foods prepared from rice anf grams. Brit. J. Nutr. 21:467-73.

248. Geddes, W.F., Hildebrand, F.C. and Anderson, J.A. 1941 The effect of wheat type, protein content, and malting conditions on the properties of malted wheat flour. Cereal Chem. 18:42-60.

249. Shands, H.L., Dickson, A.D., Dickson, J.G. and Burkhart, B.A. 1941 The influence of temperature, moisture and growth time on the malting quality of four barley varieties. Cereal Chem. 18:370-94.

250. Kneen, E., Miller, B.S. and Sandstedt, R.M. 1942 The influence of temperature on the development of amylase in germinating wheat. Cereal Chem. 19:11-27.

251. Sallans, H.R. and Anderson, J.A. 1939 Observations on the study of varietal differences in the malting quality of barley. Can. J. Res. C17:57-71.

252. Dickson, A.D. and Burkhart, B.A. 1942 Changes in the barley kernel during malting: Chemical comparisons of germ and distal portions. Cereal Chem. 19:251-62.

253. Dickson, A.D., Olson, W.J. and Shands, H.L. 1947 Amylase activity of three barley varieties as influenced by different malting conditions. Cereal Chem. 24:325-37.

254. Dickson, J.G. and Geddes, W.F. 1949 Effect of wheat class and germination moisture and time on malt yield and amylase activity of malted wheat. Cereal Chem. 26:404-14.

255. Morinaga, T. 1926 Effect of alternating temperatures upon the germination of seeds. Amer. J. Bot. 13:141-58.

256. Shands, H.L., Dickson, A.D. and Dickson, J.G. 1942 The effect of temperature change during malting on four barley varieties. Cereal Chem. 19:471-80.

257. Schimper, A.F.W. 1903 Plant geography. Eng. Translation by W.R. Fisher. Clarendon Press, Oxford.

258. Duggar, B.M. 1911 Plant Physiology. McMillan Co., New York.

259. Jones, F.R. 1921 Effect of soil temperature upon development of nodules on the roots of certain legumes. J. Agr. Res. 22:17-31.

260. Percival, J. 1921 The wheat plant. Duckworth and Co., London.

261. Atterberg, A. 1907 Die nachereife des Getreides. Landw. Vers. Stat. 67:129-143.

262. Malsbury, M.R. 1920 Imbibition and germination of wheat at different temperatures. Unpublished data. Plant Physiology Division, University of Illinois.

263. Wilson, H.K. and Hottes, C.F. 1927 Wheat germination studies with particular reference to temperature and moisture relationships. J. Amer. Soc. Agron. 19:181-90.

264. Coffman, F.A. 1923 Minimum temperature of germination of seeds. J. Am. Sco. Agron. 15:257-70.

265. Dickson, J.G. 1923 The influence of soil temperature and moisture on the germination of wheat treated with different seed treatments. Proc. Assoc. Seed Analyses of N. Amer. 14:68-72.

266. Wilson, H.K. 1928 Wheat, soybean, and oat germination studies with particular reference to temperature relationships. J. Am. Soc. Agron. 20:599-619.

267. Edwards, T.I. 1934 Relations of germinating soybeans to temperature and length of incubation time. Plant Physiol. 9:1-30.

268. Hartmann, O. 1919 Über den Einfluss der Temperatur auf Plasma, Kern und Nucleolus und cytologische Gleichgewichtszustände. Zellphysiologische Experimente an Pflanzen. Arch. Zellforschung 15:177-248.

269. True, R.H. 1914 The harmful action of distilled water. Amer. J. Bot. 1:255-73.

270. Kidd, R. and West, C. 1918 Physiological pre-determinion: The influence of the physiological condition of the seed upon the course of subsequent growth and upon the yield. 1. The effects of soaking seeds in water. Ann. Appl. Biol. 5:1-10.

271. Kidd, F. and West, C. 1919 The influence of temperature on the soaking of seeds. New Phytol. 18:35-39.

272. Gericke, W.F. 1929 Some relations of maintained temperatures to germination and the early growth of wheat in nutrient solutions. Philippine J. Sci. 38:215-38.

273. Tang, P. 1931 An experimental study of the germination of wheat seed under water as related to temperature and aeration. Plant Physiol. 6:203-48.

274. Bailey, W.M. 1933 Structural and metabolic after-effects of soaking seeds of Phaseolus. Bot. Gaz. 94:688-713.

275. Eyster, H.C. 1940 The cause of decreased germination of bean seeds soaked in water. Am. J. Bot. 27:652-59.

276. Barton, L.V. 1950 Relation of different gasses to the soaking injury of seeds. Centr. Boyce Thompson Inst. Pl. Res. 16:55-71.

277. Orphanos, P.I. and Heydecker, W. 1968 On the nature of the soaking injury of Phaseolus vulgaris seeds. J. Exp. Bot. 19:770-84.

278. Mathews, S. and Bradnock, W.T. 1968 The detection of seed samples of wrinkle-seeded peas (Pisum sativum L.) of potentially low planting value. Proc. int. Seed Test. Ass. 32:553-63.

279. Perry, D.A. and Harrison, J.G. 1970 The deleterious effect of water and low temperature on germination of pea seed. J. Exp. Bot. 21:504-12.

280. Morinaga, T. 1926 The favorable effect of reduced oxygen supply upon the germination of certain seeds. Amer. J. Bot. 13:159-66.

281. Ramakrishnan, C.V. 1949 Studies on Indian fermented foods. Baroda J. Nutr. 6:1-57.

282. Devadas, R.P. and Murthy, N.K. 1977 Nutrition of the preschool child in India. Wld. Rev. Nutr. Diet. 27:1-33.

283. Harden, A. and Zilva, S.S. 1918 An investigation of beer for antineuritic and antiscorbutic potency. J. Inst. Brew. 24:197-208.

284. Kucera, C. 1928 Variations de la teneur en vitamines B et C des graines de céréales au cours de la germination. Compt. Rend. Soc. Biol. 99:967-70.

285. Matsuoka, T. 1936 Study of vitamin C. No. 18. Effect of light on production of vitamin C. Japanese J. Agr. Chem. 12:1203-10.

286. Kucera, C. 1928 Sur la localisation des vitamines B et C dans la jeune plante. Compt. Rend. Soc. Biol. 99:971-72.

287. Von Ohlen, F.W. 1931 A microchemical study of soybeans during germination. Am. J. Bot. 18:30-49.

288. Embrey, H. and Wang, T.C. 1921 Analysis of some Chinese foods. Chin. Med. J. 35:247-57.

289. Pal, H.K. and Pal, R.K. 1945 Some observations on the synthesis of proteins by plants. Annals of Biochemistry and Experimental Med. V:131-34. Calcutta, India.

290. Hurst, C.J. and Sudia, T.W. 1973 The effect of light on the use of the nitrogen reserves of germinating soybean seeds. Amer. J. Bot. 60:1034-40.

291. Heller, V.G. 1927 Nutritive properties of the mung bean. J. Biol. Chem. 75:435-42.

292. Pomeranz, Y., Shogren, M.D. and Finney, K.F. 1977 Flour from germinated soybeans in high protein bread. J. Fd. Sci. 42:824-28, 842.

293. Wang, Y.D. and Fields, M.L. 1978 Germination of corn and sorghum in the home to improve nutritive value. J. Fd. Sci. 43:1113-15.

294. Wang, Y.D. and Fields, M.L. 1978 A research note: Enrichment of home prepared tortillas made from germinated corn. J. Fd. Sci. 43:1630-31.

295. Hamad, A.M. and Fields, M.L. 1979 Evaluation of the protein quality and available lysine of germinated and fermented cereals. J. Fd. Sci. 44:456-59.

296. Hasim, N.B. and Fields, M.L. 1979 Germination and relative nutritive value of corn meal and corn chips. J. Fd. Sci. 44:936-37.

297. Niyogi, S.P., Narayana, N. and Desai, B.G. 1931 Studies in the nutritive value of Indian vegetable food-stuffs. Part 1. Nutritive values of pigeon pea (Cajanus indicus) and field pea (Pisum arvense, Linn). Ind. J. Med. Res. 18:1217-29.

298. Niyogi, S.P., Narayana, N. and Desai, B.G. 1931 Studies in the nutritive value of Indian vegetable food-stuffs. Part II. Nutritive values of (1) Bengal gram (Cicer arietinum, Linn.) (2) Horse gram (Dolichos biflorus), (3) Lablab pea (Dolichos lablab). Ind. J. Med. Res. 19:475-83.

299. Niyogi, S.P., Narayana, N. and Desai, B.G. 1932 Studies in the nutritive value of Indian vegetable food-stuffs. Part III. Nutritive values of Lentil, Lens esculenta, Moench, Cowpea, Vigna catjang, Walp, and Aconite bean, Phaseolus aconitifolius, Jacq. Ind. J. Med. Res. 19:859-66.

300. Niyogi, S.P., Narayana, N. and Desar, B.G. 1932 Studies in the nutritive value of Indian vegetable food-stuffs. Part IV. Nutritive value of green gram, Phaseolus radiatus and blackgram, Phaseolus mungo. Ind. J. Med. Res. 19:1041-54.

301. Haffenrichter, A.L. 1928 Respiration of the soybean. Bot. Gaz. 85:271-98.

302. Edwards, T.I. 1932 Temperature relations of seed germination. Quart. Rev. Biol. 7:428-43.

303. Atterberg, A. 1928 Die Nachreife des Getreides. Landw. Versuchsst. 67:129-43.

304. Abdul-Baki, A.A. and Anderson, J.D. 1970 Viability and leaching of sugars from germinating barley. Crop. Sci. 10:31-4.

305. Mayer, A.M. and Polkjaoff-Mayber, A. 1963 The germination of seeds. Macmillan, New York, NY.

306. Hendricks, S.B. and Taylorson, R.B. 1972 Promotion of seed germination by nitrates and cyanides. Nature 237:169-70.

307. Hendricks, S.B. and Taylorson, R.B. 1974 Promotion of seed germination by nitrate, nitrite, hydroxylamine, and ammonium salts. Pl. Plysiol. 54:304-9.

308. Reynolds, T. and MacWilliam, I.C. 1966 Water uptake and enzymic activity during steeping of barley. J. Inst. Brew. 72:166-70.

309. Mentzer, C. 1940 Quelques donñees nouvelles sur l'acide ascorbique dons le règne vegétal. Societe De chimie Biologique 22:445-57.

310. Matsuoka, T. 1931 Study of vitamin C. No. 5 Vitamin C in seed sprouts when grown in dark (No. 4). Japanese J. Soc. Agr. Chem. 7:1070-81.

311. Simonik, F. 1928 Teneur en vitamine C des graines de légumineuses pendant la germination. Compt. Rend. Soc. Biol. 99:431-32.

312. Morad, M.M., Leung, H.K., Hsu, D.L. and Finney, P.L. 1980 Effect of germination on physicochemical and bread-baking properties of yellow pea, lentil, and faba bean flours and starches. Cereal Chem. 57:390-96.

312. Bates, L.S., Finney, P.L. and Jones, M. 1981 Effects of germination time and degree on dry weight loss, water imbibition and complete amino acid transformation of 12 highly devergent U.S. wheat variety composites. Paper #208, 66th Annual Meeting of the AACC, Denver, CO, Oct. 25-29, 1981.

313. Hsu, D.L., Leung, H.K., Morad, M.M., Finney, P.L. and Leung, C.T. 1982 Effect of germination on electrophoretic, functional, and bread-baking properties of yellow pea, lentil, and faba bean protein isolates. Cereal Chem. 59:344-50.

INDEX